GREATways to Teach and Learn™

# Connect with Math
## Grade 1

Written by

Patricia Pedigo and Dr. Roger DeSanti

Edited and produced by
Patricia Pedigo

©2008 Plutarch Publications, Inc. PPI -2001

ISBN 978-1-934990-05-6

**GREATways to Teach and Learn: Connect with Math Grade 1**
Published by
**Plutarch Publications, Inc.**
U.S.A.
Email (for customer service enquires): plutarch_inc@yahoo.com
Copyright© 2008 Plutarch Publications, Inc.

Copyright © 2008, Plutarch Publications, Inc., Mandeville, Louisiana. All rights reserved. The purchase of this material entitles the buyer to reproduce worksheets and activities for classroom use only—not for commercial resale. Reproduction of these materials for an entire school or district is prohibited. No part of this book may be reproduced (except as noted above), stored in a retrieval system, or transmitted in any form or by any means (mechanically, electronically, recording, etc.) without the prior written consent of Plutarch Publications, Inc.

**ISBN-13: 978-1-934990-05-6**
**ISBN-10: 1-934990-05-1**

©2008 Plutarch Publications, Inc. PPI -2001

## About the series ...

The GREATways: Instruction series are books intended to supplement textbooks. Over sixty pages of activities presented in each GREATways: Instruction book engage the learner in active practice of basic skills required at the appropriate grade level. Activities are designed with various learning styles in mind to include every child in the learning process.

Each book contains two pages of *Quick Cues,* a handy list of important vocabulary, rules, or examples of standards covered in that GREATways: Instruction book. The page "How to Use This Book" provides suggestions and ideas for using *Quick Cues* for additional instruction or practice.

GREATways: Instruction books are designed to comply with State Curriculum Standards. Although the level at which specific topics are mandated may vary from State to State, many State Curriculum Standards agree on the grade level at which most skills are introduced. The GREATways: Instruction series focuses on those standards that are commonly introduced at each grade level. The Score Computation Chart (page 4) and the Standards Competency Chart (page 5) provide a viable means to assess the level at which a child is able to complete each standard presented.

The goal of this series is to provide grade appropriate standards, practice, and application in a straight-forward, easy to understand manner. Appropriate materials and presentation produce comprehension. Practice produces proficiency. Application produces students able to interact with the real world.

## About the authors and editor .....

**Patricia Pedigo, M.Ed.** in elementary education, also earned the Reading Specialist endorsement. She has more than 20 years experience in elementary and junior high classrooms and a passion for working with "learning different" children. Patricia has authored and/or edited 50 instructional books that are used in classrooms across North America.

**Roger DeSanti Sr., Ed.D.** in Reading and Special Education, is a Professor of Education whose area of expertise is literacy and the learning process. He has over 30 years of classroom experience working with and educating children and their teachers. Roger has over 100 publications, including instructional books that are used in classrooms across North America.

# Connect With Math Grade 1

| | |
|---|---|
| How to Use This Book ..................... 1 | Word Problems ........................... 40 |
| Quick Cues Vocabulary ................. 2 | Word Problems ........................... 41 |
| Quick Cues Vocabulary ................. 3 | Word Problems ........................... 42 |
| Score Computation Chart .............. 4 | Word Problems ........................... 43 |
| Standards Competency Chart ........ 5 | Word Problems ........................... 44 |
| **Identify Numbers** | **Patterns** |
| Count/Write to Ten ......................... 6 | Count by 2 ................................... 45 |
| Count/Write to Ten ......................... 7 | Count by 2 ................................... 46 |
| Count/Write to Ten ......................... 8 | Count by 2 ................................... 47 |
| Count/Write to Ten ......................... 9 | Count by 5 ................................... 48 |
| Count/Write to Twenty ................... 10 | Count by 5 ................................... 49 |
| Count/Write to Twenty ................... 11 | Count by 5 ................................... 50 |
| Count/Write to Twenty ................... 12 | Count by 10 ................................. 51 |
| Count/Write to Twenty ................... 13 | Count by 10 ................................. 52 |
| Count/Write to Thirty ..................... 14 | Count by 10 ................................. 53 |
| Count/Write to Thirty ..................... 15 | **Measurement** |
| Count/Write to Thirty ..................... 16 | Time Terms .................................. 54 |
| Count/Write to Thirty ..................... 17 | Time Terms .................................. 55 |
| **Addition/Subtraction Facts** | Time to the Hour .......................... 56 |
| Add Facts to 10 ............................ 18 | Time to the Half Hour ................... 57 |
| Add Facts to 10 ............................ 19 | Time to the Half Hour ................... 58 |
| Subtraction Facts to 10 ................. 20 | Counting Money ........................... 59 |
| Subtraction Facts to 10 ................. 21 | Counting Money ........................... 60 |
| Add/Subtract Facts to 10 .............. 22 | Counting Money ........................... 61 |
| Add/Subtract Facts to 10 .............. 23 | Inches/Feet ................................. 62 |
| Add/Subtract Facts to 10 .............. 24 | Inches/Feet ................................. 63 |
| Add Facts to 20 ............................ 25 | **Geometry - shapes** |
| Add Facts to 20 ............................ 26 | Fractions ..................................... 64 |
| Subtraction Facts to 20 ................. 27 | Fractions ..................................... 65 |
| Subtraction Facts to 20 ................. 28 | Fractions ..................................... 66 |
| Add/Subtract Facts to 20 .............. 29 | Identify Shapes ............................ 67 |
| Add/Subtract Facts to 20 .............. 30 | Identify Shapes ............................ 68 |
| Add/Subtract Facts to 20 .............. 31 | Construct Shapes ........................ 69 |
| Order Sets ................................... 32 | **Answer Keys** |
| Order Sets ................................... 33 | Pages 6 - 9 ..... 70   Pages 38 - 41 ... 78 |
| Order Sets ................................... 34 | Pages 10 - 13 .. 71   Pages 42 - 45 ... 79 |
| **Story Problems** | Pages 14 - 17 .. 72   Pages 46 - 49 ... 80 |
| Word Problems ........................... 35 | Pages 18 - 21 .. 73   Pages 50 - 53 ... 81 |
| Word Problems ........................... 36 | Pages 22 - 25 .. 74   Pages 54 - 57 ... 82 |
| Word Problems ........................... 37 | Pages 26 - 29 .. 75   Pages 58 - 61 ... 83 |
| Word Problems ........................... 38 | Pages 30 - 33 .. 76   Pages 62 - 65 ... 84 |
| Word Problems ........................... 39 | Pages 34 - 37 .. 77   Pages 66 - 69 ... 85 |

©2008 Plutarch Publications, Inc. PPI -2001

# How to use this book ...

GREATways: Instruction books offer several features designed to enhance the learning process and assist the teacher in assessing the learner's progress. On the next few pages you will find Quick Cues, a Score Computation Chart, a Standards Competency Chart, and recommendations based on the competency level of the learner.

*QUICK CUES:* This book includes two pages of *Quick Cues* which are important facts at your fingertips. The Quick Cues found on page two and three of this book lists addition facts and math words with their definitions, all of which should be part of the basic math vocabulary of first grade learners. Ways to use these pages are as varied as the number of learners, but here are a few suggestions to get started:
- Have the learner make flashcards of the addition facts and math words with the answers or definitions written on the back. These may be used in games and drills to review and commit the information to memory.
- Learners may practice fractions by cutting items such as food or paper into halves and fourths.
- Have learners keep a list of things they do in the morning, afternoon, evening, and night.
- Learners may list important future events (such as family birthdays) and use a calendar to find the days for those events.
- Have learners start a coin jar or piggy bank. Count the coins weekly.

**SCORE COMPUTATION CHART:** This assessment tool can be found on page four of this book. After the learner completes an activity in this book, record the number of correct items on the score computation chart. When all pages for a listed standard have been completed, tally the number of correct answers and record it in the column on the far right (under the total of correct answers possible). Transfer the learner's totals to the chart on page five to find the level of competency.

**STANDARDS COMPETENCY CHART:** Use the total number correct scores from page four to identify the level at which the learner comprehends/applies the standard. The range of scores within each level (Mastery, Instructional, Basic, and Limited) are approximate indicators of how well the learner understands and can apply each standard. The degree of competency at that level will vary with the score. For example, a score of 38 in Telling Time indicates Mastery, but is close to Instructional and the learner could benefit from more practice with that standard. Recommendations based on the competency level are offered at the bottom of the page.

# Quick Cues

## Addition Facts to Twenty

| | | | | |
|---|---|---|---|---|
| 1 + 0 = 1 | 2 + 0 = 2 | 3 + 0 = 3 | 4 + 0 = 4 | 5 + 0 = 5 |
| 1 + 1 = 2 | 2 + 1 = 3 | 3 + 1 = 4 | 4 + 1 = 5 | 5 + 1 = 6 |
| 1 + 2 = 3 | 2 + 2 = 4 | 3 + 2 = 5 | 4 + 2 = 6 | 5 + 2 = 7 |
| 1 + 3 = 4 | 2 + 3 = 5 | 3 + 3 = 6 | 4 + 3 = 7 | 5 + 3 = 8 |
| 1 + 4 = 5 | 2 + 4 = 6 | 3 + 4 = 7 | 4 + 4 = 8 | 5 + 4 = 9 |
| 1 + 5 = 6 | 2 + 5 = 7 | 3 + 5 = 8 | 4 + 5 = 9 | 5 + 5 = 10 |
| 1 + 6 = 7 | 2 + 6 = 8 | 3 + 6 = 9 | 4 + 6 = 10 | 5 + 6 = 11 |
| 1 + 7 = 8 | 2 + 7 = 9 | 3 + 7 = 10 | 4 + 7 = 11 | 5 + 7 = 12 |
| 1 + 8 = 9 | 2 + 8 = 10 | 3 + 8 = 11 | 4 + 8 = 12 | 5 + 8 = 13 |
| 1 + 9 = 10 | 2 + 9 = 11 | 3 + 9 = 12 | 4 + 9 = 13 | 5 + 9 = 14 |
| 1 + 10 = 11 | 2 + 10 = 12 | 3 + 10 = 13 | 4 + 10 = 14 | 5 + 10 = 15 |
| 6 + 0 = 6 | 7 + 0 = 7 | 8 + 0 = 8 | 9 + 0 = 9 | 10 + 0 = 10 |
| 6 + 1 = 7 | 7 + 1 = 8 | 8 + 1 = 9 | 9 + 1 = 10 | 10 + 1 = 11 |
| 6 + 2 = 8 | 7 + 2 = 9 | 8 + 2 = 10 | 9 + 2 = 11 | 10 + 2 = 12 |
| 6 + 3 = 9 | 7 + 3 = 10 | 8 + 3 = 11 | 9 + 3 = 12 | 10 + 3 = 13 |
| 6 + 4 = 10 | 7 + 4 = 11 | 8 + 4 = 12 | 9 + 4 = 13 | 10 + 4 = 14 |
| 6 + 5 = 11 | 7 + 5 = 12 | 8 + 5 = 13 | 9 + 5 = 14 | 10 + 5 = 15 |
| 6 + 6 = 12 | 7 + 6 = 13 | 8 + 6 = 14 | 9 + 6 = 15 | 10 + 6 = 16 |
| 6 + 7 = 13 | 7 + 7 = 14 | 8 + 7 = 15 | 9 + 7 = 16 | 10 + 7 = 17 |
| 6 + 8 = 14 | 7 + 8 = 15 | 8 + 8 = 16 | 9 + 8 = 17 | 10 + 8 = 18 |
| 6 + 9 = 15 | 7 + 9 = 16 | 8 + 9 = 17 | 9 + 9 = 18 | 10 + 9 = 19 |
| 6 + 10 = 16 | 7 + 10 = 17 | 8 + 10 = 18 | 9 + 10 = 19 | 10+ 10 = 20 |

©2008 Plutarch Publications, Inc.  PPI -2001

# Quick Cues

## Words in Math

### NUMBERS

| one | eleven | twenty-one |
| two | twelve | twenty-two |
| three | thirteen | twenty-three |
| four | fourteen | twenty-four |
| five | fifteen | twenty-five |
| six | sixteen | twenty-six |
| seven | seventeen | twenty-seven |
| eight | eighteen | twenty-eight |
| nine | nineteen | twenty-nine |
| ten | twenty | thirty |

### FRACTIONS

$\frac{1}{2}$  $\frac{1}{2}$

each part equals 1 out of two pieces or 1/2 of the apple

$\frac{1}{4}$ $\frac{1}{4}$
$\frac{1}{4}$ $\frac{1}{4}$

each part equals 1 out of four pieces or 1/4 of the sandwich

### TIME OF DAY

**day** - 24 hours from midnight until the next midnight
**daylight** - the time from sunrise until sunset
**morning** - the time from sunrise until 12 o'clock noon
**noon** - 12 o'clock in the middle of the day
**afternoon** - the time from noon until 6 o'clock.
**evening** - near the end of the day, around sunset
**night** - the time when it is dark outside
**midnight** - 12 o'clock at night, the middle of the night

### MEASUREMENT

Use **INCHES** to measure small things like buttons.
Use **FEET** to measure bigger things like walls.
Use **YARDS** to measure large things like buildings.
Use **MILES** to measure long distances.

### MONEY

**penny** = 1 ¢
**nickel** = 5 ¢
**dime** = 10 ¢
**quarter** = 25 ¢

# Score Computation Chart

## Connect with Math
## Grade 1

| Count/Write to 30 | | | | | | | | | | | | | Score |
|---|---|---|---|---|---|---|---|---|---|---|---|---|---|
| Page number | 6 | 7 | 8 | 9 | 10 | 11 | 12 | 13 | 14 | 15 | 16 | 17 | |
| # possible | 10 | 10 | 10 | 10 | 10 | 10 | 10 | 10 | 10 | 10 | 10 | 10 | 120 |
| # correct | | | | | | | | | | | | | |
| **Facts to 20** | | | | | | | | | | | | | |
| Page number | 18 | 19 | 20 | 21 | 22 | 23 | 24 | | | | | | |
| # possible | 10 | 10 | 10 | 10 | 12 | 12 | 24 | | | | | | 88 |
| # correct | | | | | | | | | | | | | |
| **Order Sets** | | | | | | | | | | | | | |
| Page number | 32 | 33 | 34 | | | | | | | | | | |
| # possible | 30 | 40 | 30 | | | | | | | | | | 100 |
| # correct | | | | | | | | | | | | | |
| **Word Problems** | | | | | | | | | | | | | |
| Page number | 35 | 36 | 37 | 38 | 39 | 40 | 41 | 42 | 43 | | | | |
| # possible | 10 | 10 | 10 | 10 | 10 | 12 | 12 | 12 | 12 | | | | 98 |
| # correct | | | | | | | | | | | | | |
| **Count by 2, 5, 10** | | | | | | | | | | | | | |
| Page number | 44 | 45 | 46 | 47 | 48 | 49 | 50 | 51 | 52 | 53 | | | |
| # possible | 16 | 16 | 16 | 10 | 16 | 16 | 10 | 16 | 16 | 8 | | | 140 |
| # correct | | | | | | | | | | | | | |
| **Telling Time** | | | | | | | | | | | | | |
| Page number | 54 | 55 | 56 | 57 | 58 | | | | | | | | |
| # possible | 6 | 6 | 10 | 10 | 10 | | | | | | | | 42 |
| # correct | | | | | | | | | | | | | |
| **Counting Money** | | | | | | | | | | | | | |
| Page number | 59 | 60 | 61 | | | | | | | | | | |
| # possible | 8 | 8 | 12 | | | | | | | | | | 28 |
| # correct | | | | | | | | | | | | | |
| **Inches and Feet** | | | | | | | | | | | | | |
| Page number | 62 | 63 | | | | | | | | | | | |
| # possible | 24 | 12 | | | | | | | | | | | 36 |
| # correct | | | | | | | | | | | | | |
| **Fractions 1/2 1/4** | | | | | | | | | | | | | |
| Page number | 64 | 65 | 66 | | | | | | | | | | |
| # possible | 24 | 24 | 12 | | | | | | | | | | 60 |
| # correct | | | | | | | | | | | | | |
| **Geometry Shapes** | | | | | | | | | | | | | |
| Page number | 67 | 68 | 69 | | | | | | | | | | |
| # possible | 28 | 28 | 10 | | | | | | | | | | 66 |
| # correct | | | | | | | | | | | | | |

©2008 Plutarch Publications, Inc. PPI-2001

# Standards Competency Chart

**Step 1:** After the learner completes each page, record the number correct on the Score Computation Chart (page 4). Calculate the total number correct for each standard.

**Step 2:** Find the learner's score for each standard in the boxes of that row. Mark the box with an X (or the learner's score) to identify the level of competency for that standard. For example, a score of 86 for the standard of Word Problems places the child on the "Instructional" level and a score of 90 would indicate the "Mastery" level.

**Step 3:** Follow the recommendation guidelines at the bottom of this page.

| Standard | Mastery | Instructional | Basic | Limited |
| --- | --- | --- | --- | --- |
| Count/Write to 30 | 120 - 108 | 107 - 90 | 89 - 72 | 71 or below |
| Facts to 20 | 88 - 79 | 78 - 66 | 65 - 53 | 52 or below |
| Order Sets | 100 - 90 | 89 - 75 | 74 - 60 | 59 or below |
| Word Problems | 98 - 88 | 87 - 74 | 73 - 59 | 58 or below |
| Count by 2, 5, 10 | 140 - 126 | 125 - 105 | 104 - 84 | 83 or below |
| Telling Time | 42 - 38 | 37 - 32 | 31 - 25 | 24 or below |
| Counting Money | 28 - 25 | 24 - 23 | 22 - 21 | 20 or below |
| Inches and Feet | 36 - 32 | 31 - 27 | 26 - 22 | 21 or below |
| Fractions 1/2 1/4 | 60 - 54 | 53 - 45 | 44 - 36 | 35 or below |
| Geometry Shapes | 66 - 59 | 58 - 49 | 48 - 39 | 38 or below |

## Recommendation Guidelines

**Mastery:** The learner is capable of using this standard independently. Move on to the next higher grade level.

**Instructional:** The learner has a working understanding of the standard, but needs some guided practice on this grade level.

**Basic:** The learner has minimal grasp of the standard and needs direct instruction and guided practice to apply the concept fully. The learner could benefit from moving one grade level lower for review and extra practice before approaching the standard at this level once again.

**Limited:** The learner has a limited understanding of the standard and should be moved to the next lower grade level for instruction and practice.

©2008 Plutarch Publications, Inc. PPI -2001

Name _____   Standard: Numbers/Words

Count the balls in each picture. Write the number and the number word.

| | | |
|---|---|---|
| ◯ | 1 | one |
| ◯◯ | 2 | two |
| ◯◯◯ | 3 | three |
| ◯◯◯◯ | 4 | four |
| ◯◯◯◯◯ | 5 | five |
| ◯◯◯◯ ◯ | 6 | six |
| ◯◯◯◯ ◯◯ | 7 | seven |
| ◯◯◯◯ ◯◯◯ | 8 | eight |
| ◯◯◯◯ ◯◯◯◯◯ | 9 | nine |
| ◯◯◯◯ ◯◯◯◯◯ | 10 | ten |

Name _____  Standard: Numbers/Words

Count the balls in each picture. Write the number and the number word.

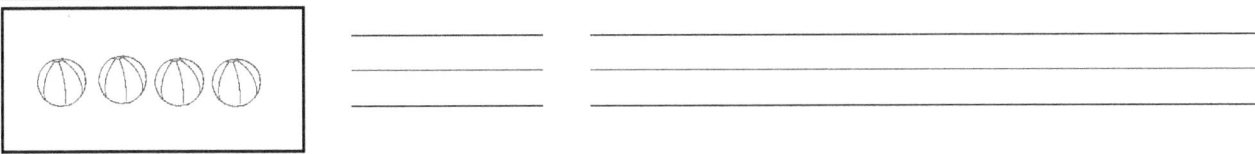

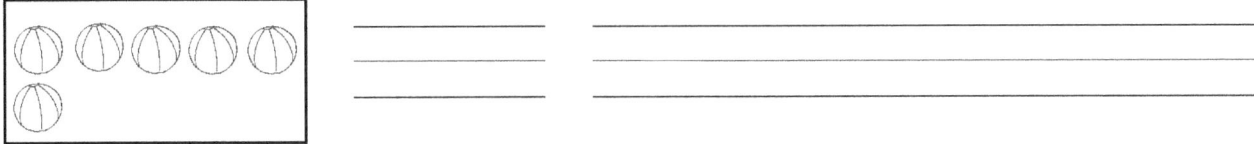

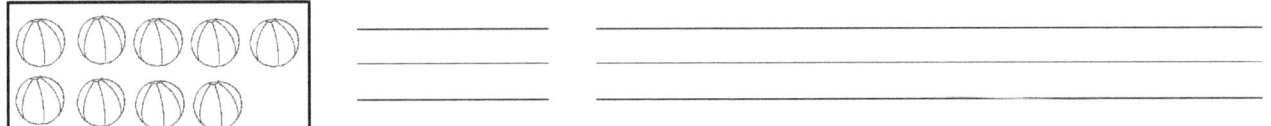

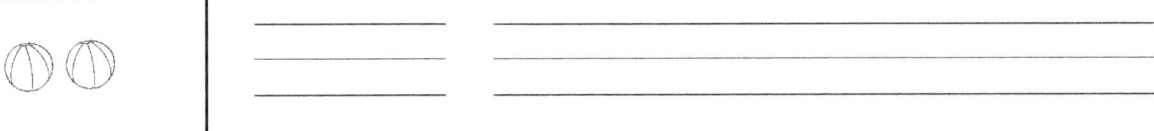

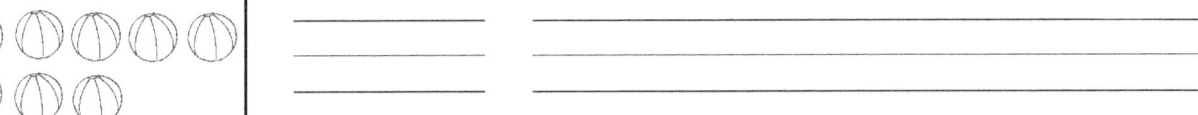

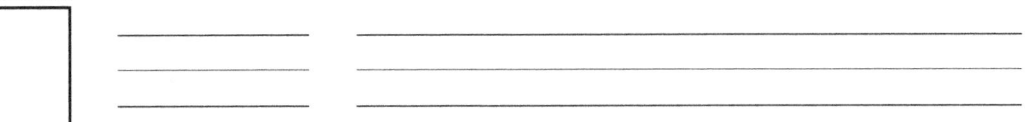

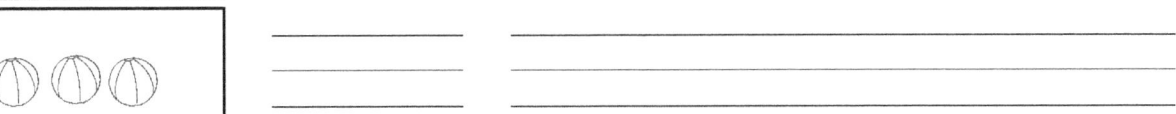

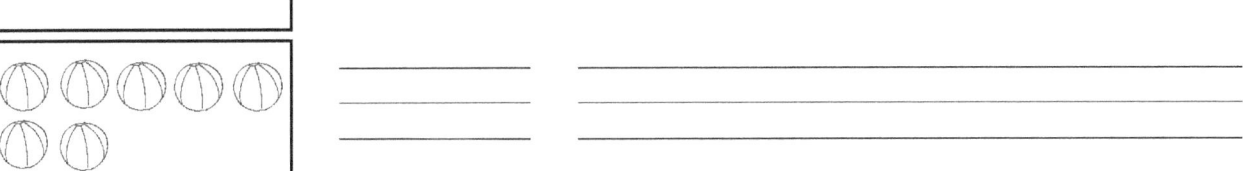

Name _____   Standard: Numbers/Words

Count the boxes in each picture. Write the number and the number word.

Name _____    Standard: Numbers/Words

Count the items in this box to answer the questions below.

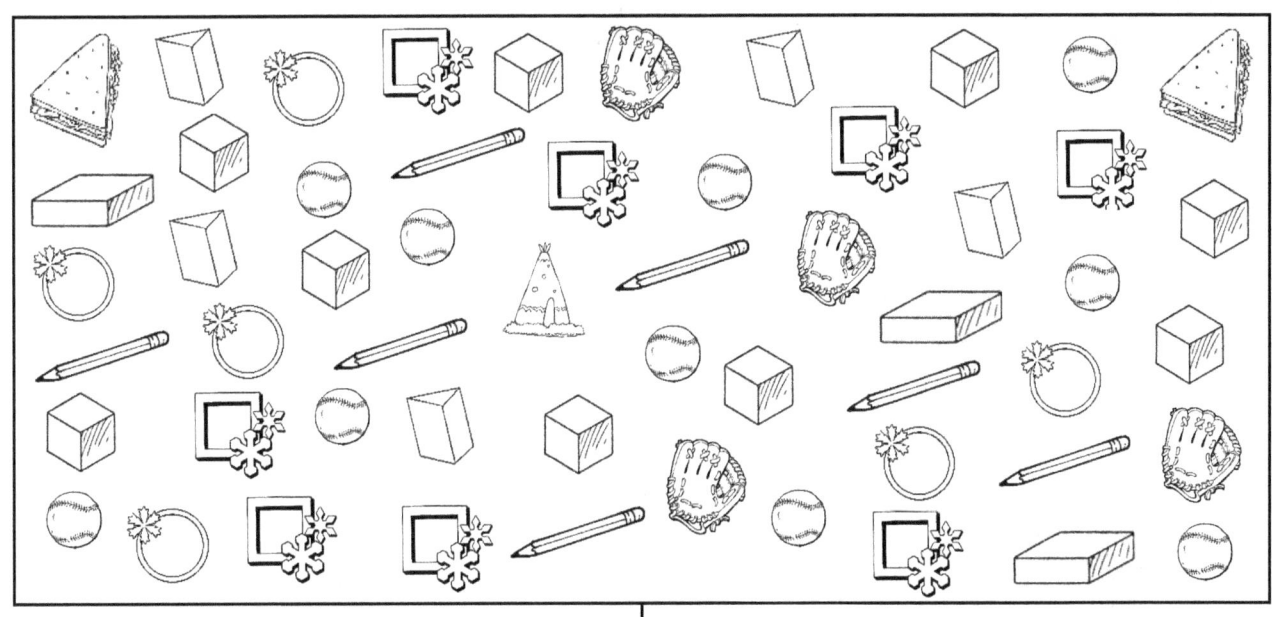

1. How many 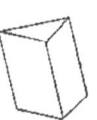 ?
_____

3. How many  ?
_____

4. How many  ?
_____

5. How many  ?
_____

6. How many  ?
_____

2. How many 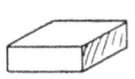 ?
_____

4. How many  ?
_____

6. How many  ?
_____

8. How many  ?
_____

10. How many  ?
_____

©2008 Plurarch Publications, Inc. PPI - 2001

Name _____  Standard: Numbers/Words

Count the kites in each picture. Write the number and the number word.

| | | |
|---|---|---|
| [5] [5] [1] | 11 | eleven |
| [5] [5] [2] | 12 | twelve |
| [5] [5] [3] | 13 | thirteen |
| [5] [5] [4] | 14 | fourteen |
| [5] [5] [5] | 15 | fifteen |
| [5] [5] [5] [1] | 16 | sixteen |
| [5] [5] [5] [2] | 17 | seventeen |
| [5] [5] [5] [3] | 18 | eighteen |
| [5] [5] [5] [4] | 19 | nineteen |
| [5] [5] [5] [5] | 20 | twenty |

©2008 Plurarch Publications, Inc. PPI - 2001

Name _____   Standard: Numbers/Words

Count the kites in each picture. Write the number and the number word.

Name _____  Standard: Numbers/Words

Count the items in each picture. Write the number and the number word.

Name _____   Standard: Number Words

Look at the items in the box. Write the number of each picture below the question.

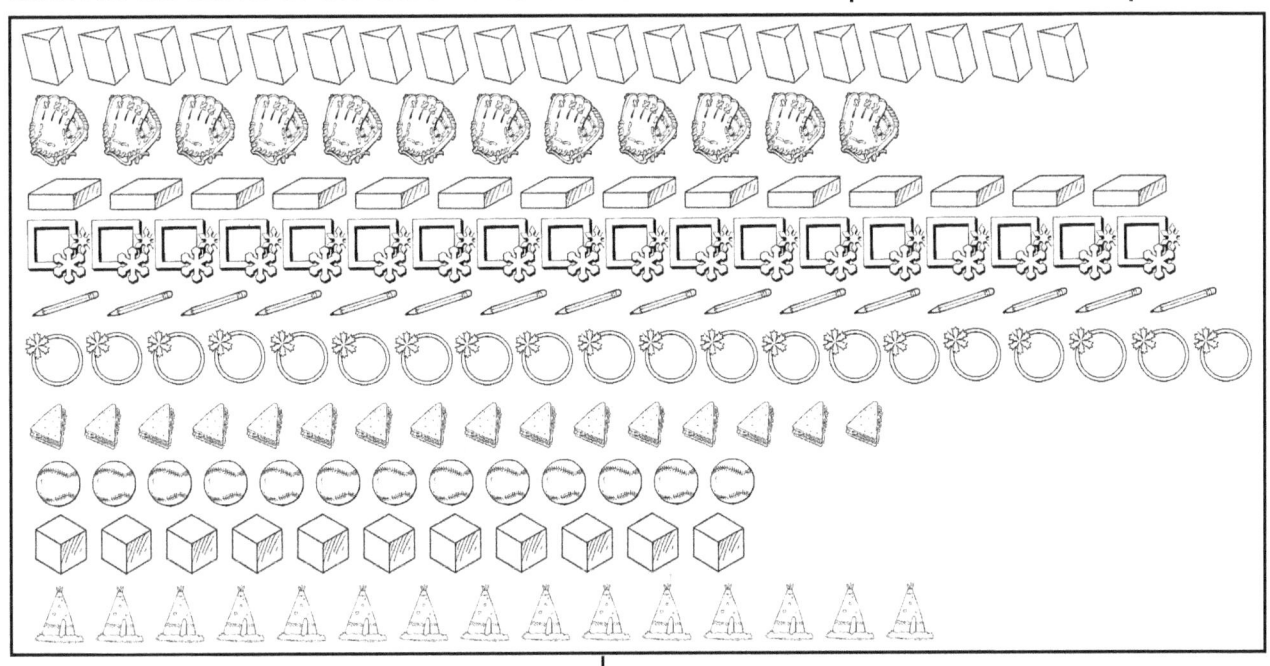

1. How many ▰ ?
_____

2. How many ▭ ?
_____

3. How many 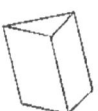 ?
_____

4. How many  ?
_____

5. How many  ?
_____

6. How many  ?
_____

7. How many  ?
_____

8. How many  ?
_____

9. How many  ?
_____

10. How many  ?
_____

©2008 Plurarch Publications, Inc. PPI - 2001

Name _____  Standard: Numbers/Words

Count the fingers in each picture. Write the number and the number word.

| | | |
|---|---|---|
| | 21 | twenty-one |
| | 22 | twenty-two |
| | 23 | twenty-three |
| | 24 | twenty-four |
| | 25 | twenty-five |
| | 26 | twenty-six |
| | 27 | twenty-seven |
| | 28 | twenty-eight |
| | 29 | twenty-nine |
| | 30 | thirty |

©2008 Plurarch Publications, Inc. PPI - 2001

Name _____  Standard: Numbers/Words

Count the fingers in each picture. Write the number and the number word.

Name _____  Standard: Numbers/Words

Count the fingers in each box. Write the number on the line.

1. How many fingers?

   5  5  5  5
      3                              _____

2. How many fingers?

   5  5  5  5
   5  5                              _____

3. How many fingers?

   5  5  5  5
   5  4                              _____

4. How many fingers?

   5  5  5  5
      5                              _____

5. How many fingers?

   5  5  5  5
   5  3                              _____

6. How many fingers?

   5  5  5  5
      1                              _____

7. How many fingers?

   5  5  5  5
      2                              _____

8. How many fingers?

   5  5  5  5
   5  1                              _____

9. How many fingers?

   5  5  5  5
   5  2                              _____

10. How many fingers?

    5  5  5  5
       4                             _____

©2008 Plurarch Publications, Inc. PPI - 2001

Name _____   Standard: Number Words

Count the books in each box. Write the number on the line.

1. How many 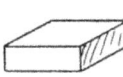 ?

_____

2. How many  ?

_____

3. How many ?

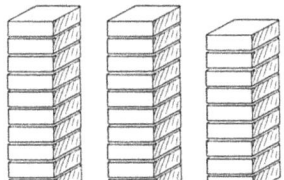
_____

4. How many  ?

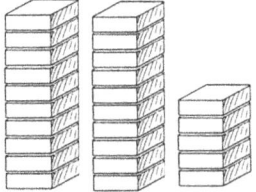
_____

5. How many ?

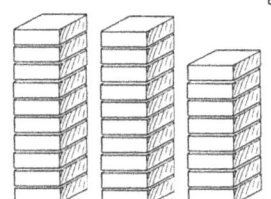
_____

6. How many ?

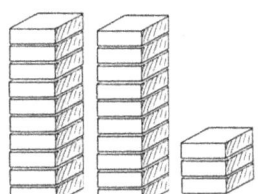
_____

5. How many ?

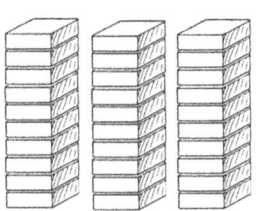
_____

5. How many ?

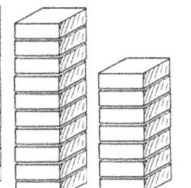
_____

5. How many ?

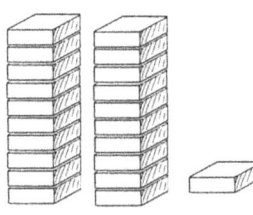
_____

5. How many ?

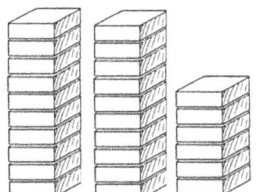
_____

©2008 Plurarch Publications, Inc. PPI - 2001

Name _____    Standard: Facts to 10

Add the items in each box. Write their total number on the line.

| | |
|---|---|
| 2 + 6 = ☐ 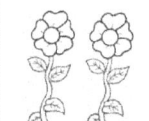 | 5 + 1 = ☐ |
| 3 + 4 = ☐ | 3 + 2 = ☐  |
| 7 + 2 = ☐  | 5 + 3 = ☐  |
|   |  |
| 6 + 1 = ☐   | 5 + 5 = ☐ 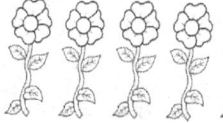  |
| 4 + 5 = ☐ 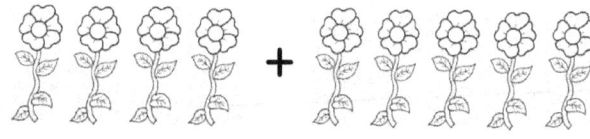 | 3 + 6 = ☐ |

©2008 Plurarch Publications, Inc. PPI - 2001        18

Name _____  Standard: Facts to 10

Add the items in each box. Write their total number on the line.

| 0 + 7 = ☐ | 8 + 2 = ☐ |
| 6 + 4 = ☐ | 3 + 6 = ☐ |
| 1 + 8 = ☐ | 4 + 4 = ☐ |
| 2 + 6 = ☐ | 7 + 3 = ☐ |
| 5 + 2 = ☐ | 3 + 3 = ☐ |

©2008 Plurarch Publications, Inc. PPI - 2001

Name _____  Standard: Facts to 10

Subtract the items in each box. Write the difference (what is left) on the line.

6 − 5 = [ 1 ]

7 − 1 = [ ]

3 − 0 = [ ]

5 − 3 = [ ]

8 − 4 = [ ]
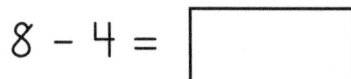

9 − 5 = [ ]

10 − 7 = [ ]

7 − 6 = [ ]

6 − 4 = [ ]

8 − 5 = [ ]
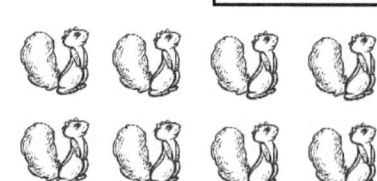

Name _____     Standard: Facts to 10

Subtract the items in each box. Write the difference (what is left) on the line.

| 10 − 4 = ☐ | 9 − 8 = ☐ |
|---|---|
| 6 − 6 = ☐ | 7 − 5 = ☐ |
| 10 − 2 = ☐ | 9 − 3 = ☐ |
| 10 − 5 = ☐ | 7 − 4 = ☐ |
| 8 − 3 = ☐ | 10 − 8 = ☐ |

Name _____   Standard: Facts to 10

Add or subtract the items then write the answer in the box below each problem.

| | |
|---|---|
| 1. $\quad 2$<br>$\underline{+2}$<br>☐ ☆  +  ☆<br>☆   ☆ | 2. $\quad 3$<br>$\underline{+4}$<br>☐ ☆☆  +  ☆☆<br>☆   ☆☆ |
| 3. $\quad 1$<br>$\underline{+5}$<br>☐ ☆  +  ☆☆☆<br>☆☆ | 4. $\quad 4$<br>$\underline{-1}$<br>☐ ☆ ☆ ☆ ☆ |
| 5. $\quad 6$<br>$\underline{-2}$<br>☐ ☆ ☆ ☆ ☆ ☆ ☆ | 6. $\quad 3$<br>$\underline{+2}$<br>☐ ☆☆  +  ☆<br>☆   ☆ |
| 7. $\quad 5$<br>$\underline{+1}$<br>☐ ☆☆☆  +  ☆<br>☆☆ | 8. $\quad 4$<br>$\underline{-3}$<br>☐ ☆ ☆ ☆ ☆ |
| 9. $\quad 5$<br>$\underline{+2}$<br>☐ ☆☆☆  +  ☆<br>☆☆   ☆ | 10. $\quad 0$<br>$\underline{+8}$<br>☐  +  ☆☆☆☆<br>☆☆☆☆ |
| 11. $\quad 7$<br>$\underline{-4}$<br>☐ ☆☆☆☆<br>☆☆☆ | 12. $\quad 8$<br>$\underline{-1}$<br>☐ ☆☆☆☆<br>☆☆☆☆ |

©2008 Plurarch Publications, Inc. PPI - 2001

Name _____  Standard: Facts to 10

Add or subtract the items then write the answer in the box below each problem.

1.  2
   − 0

2.  3
   + 7

3.  4
   + 0

4.  5
   − 5

5.  6
   + 2

6.  9
   − 2

7.  9
   − 4

8.  7
   − 0

9.  5
   + 5

10. 4
   + 3

11. 10
   − 3

12. 6
   + 4

©2008 Plurarch Publications, Inc. PPI - 2001

Name _____    Standard: Facts to 10

Add or subtract the items then write the answer below each problem.

| 1. $\begin{array}{r} 8 \\ +2 \\ \hline \end{array}$ | 2. $\begin{array}{r} 1 \\ +9 \\ \hline \end{array}$ | 3. $\begin{array}{r} 9 \\ -3 \\ \hline \end{array}$ | 4. $\begin{array}{r} 6 \\ -2 \\ \hline \end{array}$ |
|---|---|---|---|
| 5. $\begin{array}{r} 7 \\ -4 \\ \hline \end{array}$ | 6. $\begin{array}{r} 8 \\ -3 \\ \hline \end{array}$ | 7. $\begin{array}{r} 7 \\ +2 \\ \hline \end{array}$ | 8. $\begin{array}{r} 2 \\ +3 \\ \hline \end{array}$ |
| 9. $\begin{array}{r} 4 \\ +5 \\ \hline \end{array}$ | 10. $\begin{array}{r} 6 \\ +1 \\ \hline \end{array}$ | 11. $\begin{array}{r} 10 \\ -1 \\ \hline \end{array}$ | 12. $\begin{array}{r} 5 \\ -4 \\ \hline \end{array}$ |
| 13. $\begin{array}{r} 8 \\ -3 \\ \hline \end{array}$ | 14. $\begin{array}{r} 4 \\ -0 \\ \hline \end{array}$ | 15. $\begin{array}{r} 5 \\ +2 \\ \hline \end{array}$ | 16. $\begin{array}{r} 4 \\ +4 \\ \hline \end{array}$ |
| 17. $\begin{array}{r} 8 \\ -2 \\ \hline \end{array}$ | 18. $\begin{array}{r} 10 \\ -6 \\ \hline \end{array}$ | 19. $\begin{array}{r} 9 \\ +0 \\ \hline \end{array}$ | 20. $\begin{array}{r} 3 \\ +7 \\ \hline \end{array}$ |
| 21. $\begin{array}{r} 6 \\ +2 \\ \hline \end{array}$ | 22. $\begin{array}{r} 4 \\ +6 \\ \hline \end{array}$ | 23. $\begin{array}{r} 10 \\ -1 \\ \hline \end{array}$ | 24. $\begin{array}{r} 7 \\ -2 \\ \hline \end{array}$ |

©2008 Plurarch Publications, Inc. PPI - 2001

Name _____    Standard: Facts to 20

Add the items in each group. Write their total number in the box.

| 10 + 3 = ☐ | 6 + 8 = ☐ |
| 8 + 4 = ☐ | 7 + 7 = ☐ |
| 9 + 5 = ☐ | 4 + 10 = ☐ |
| 11 + 2 = ☐ | 9 + 8 = ☐ |
| 6 + 6 = ☐ | 6 + 10 = ☐ |

©2008 Plurarch Publications, Inc. PPI - 2001

Name _____   Standard: Facts to 20

Add the items in each group. Write their total number in the box.

12 + 3 = ☐

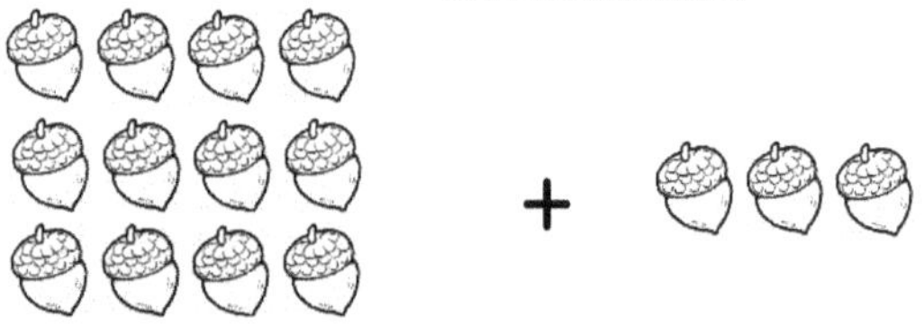

14 + 4 = ☐

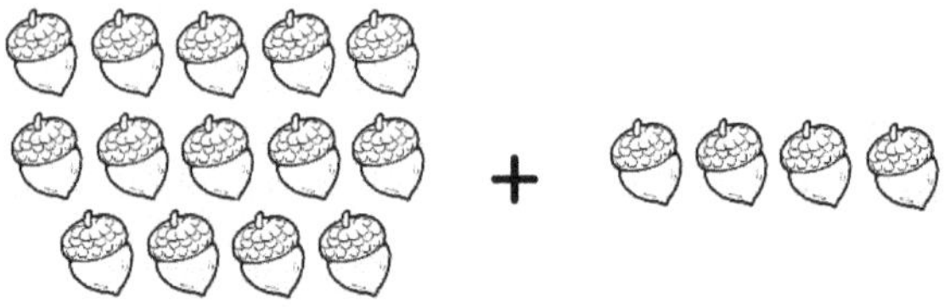

9 + 9 = ☐

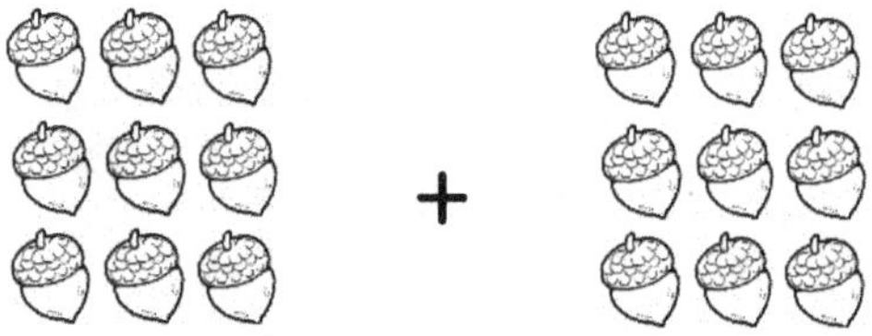

11 + 7 = ☐

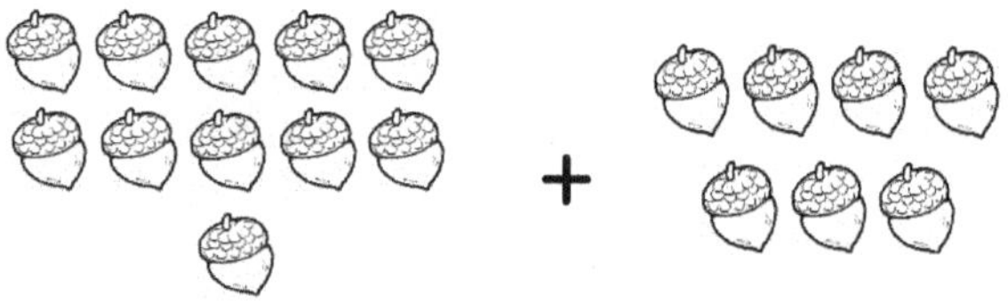

8 + 6 = ☐

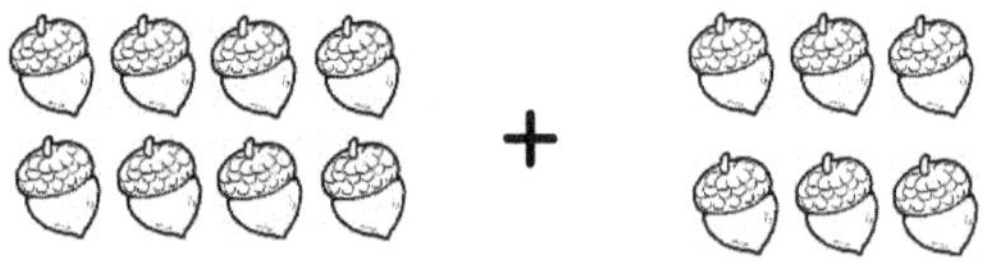

10 + 5 = ☐

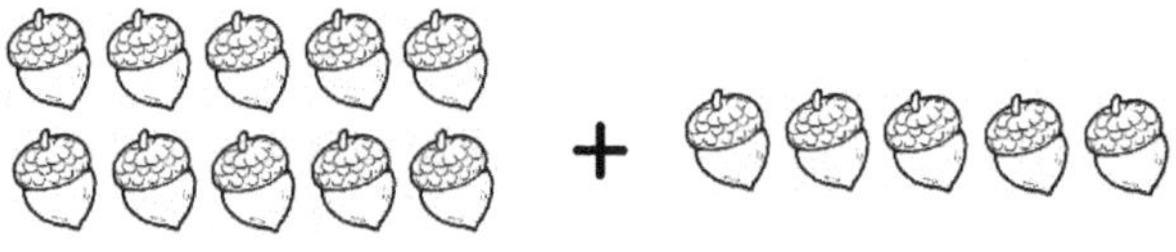

4 + 13 = ☐

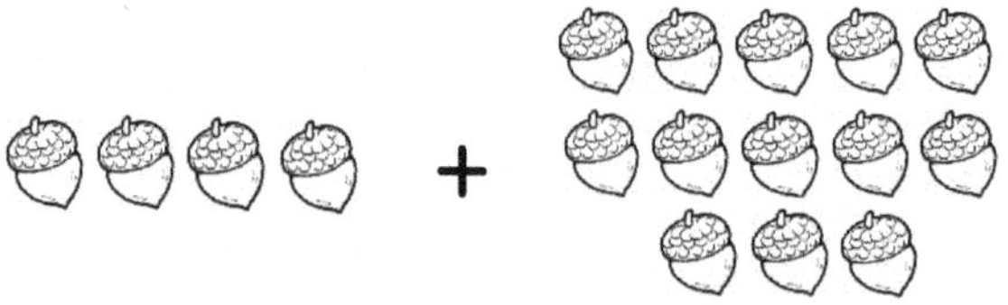

1 + 17 = ☐

9 + 10 = ☐

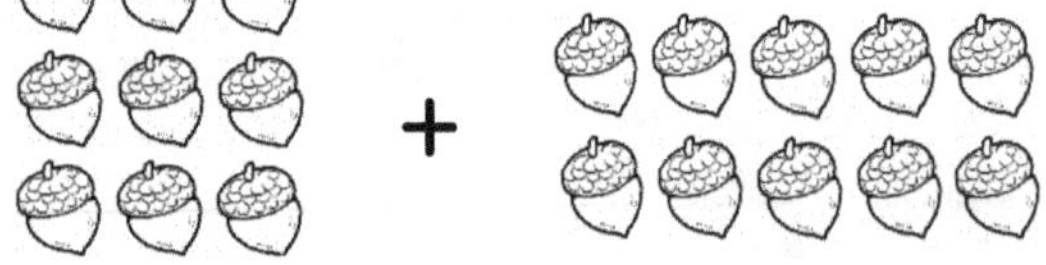

10 + 10 = ☐

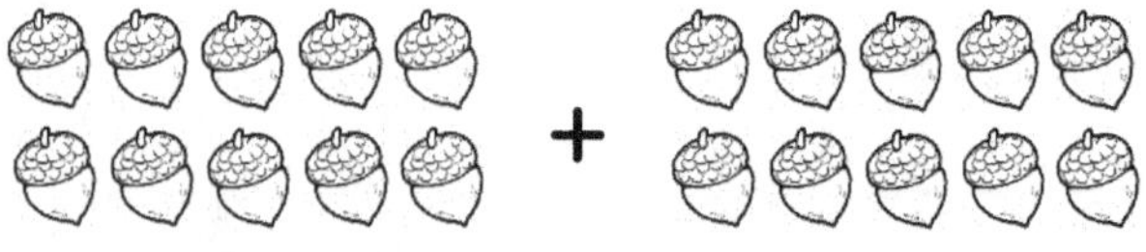

Name _____   Standard: Facts to 20

Subtract the items in each box. Write their difference in the answer box.

| 11 − 3 = ☐ | 14 − 7 = ☐ |
|---|---|
| 12 − 6 = ☐ | 15 − 8 = ☐ |
| 13 − 4 = ☐ | 11 − 9 = ☐ |
| 12 − 4 = ☐ | 11 − 4 = ☐ |
| 17 − 7 = ☐ | 18 − 9 = ☐ |

Name _____   Standard: Facts to 20

Subtract the items in each box. Write their difference in the answer box.

| 16 − 2 = ☐ | 17 − 5 = ☐ |
|---|---|
|  |  |
| 19 − 6 = ☐ | 18 − 3 = ☐ |
| (trees) |  |
| 16 − 8 = ☐ | 20 − 10 = ☐ |
|  |  |
| 20 − 6 = ☐ | 15 − 7 = ☐ |
|  |  |
| 14 − 9 = ☐ | 20 − 3 = ☐ |
|  |  |

©2008 Plurarch Publications, Inc. PPI - 2001

Name _____   Standard: Facts to 20

Add or subtract the items then write the answer in the box below each problem.

1. 13 − 8
2. 10 + 7
3. 12 − 9
4. 8 + 8
5. 9 + 5
6. 11 − 1
7. 7 + 9
8. 12 − 4
9. 13 + 2
10. 11 + 5
11. 14 − 8
12. 15 − 4

©2008 Plurarch Publications, Inc. PPI - 2001

Name _____     Standard: Facts to 20

Add or subtract the items then write the answer in the box below each problem.

1.  12
   + 6
   ☐

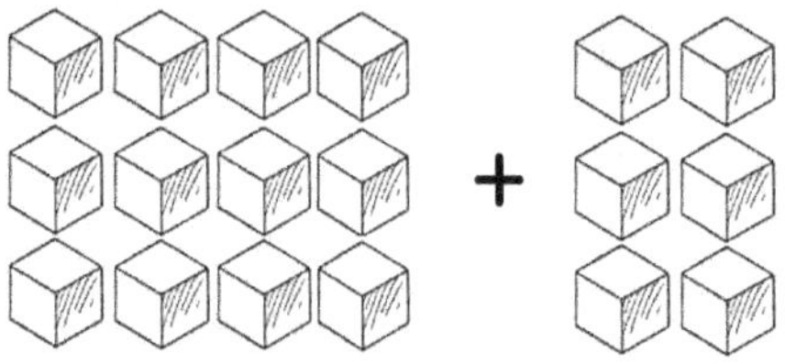

2.   9
   + 10
   ☐

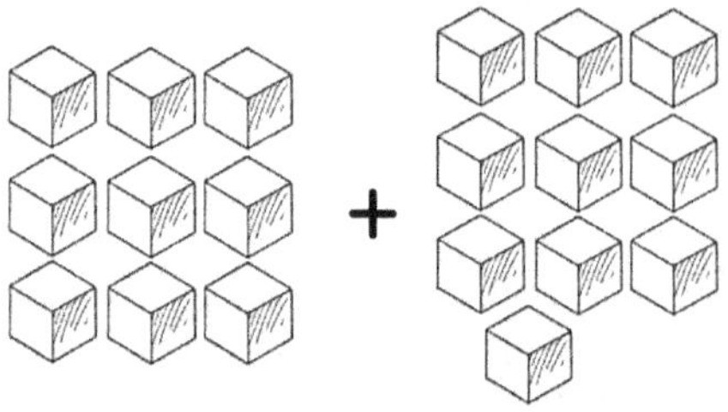

3.  13
   - 7
   ☐

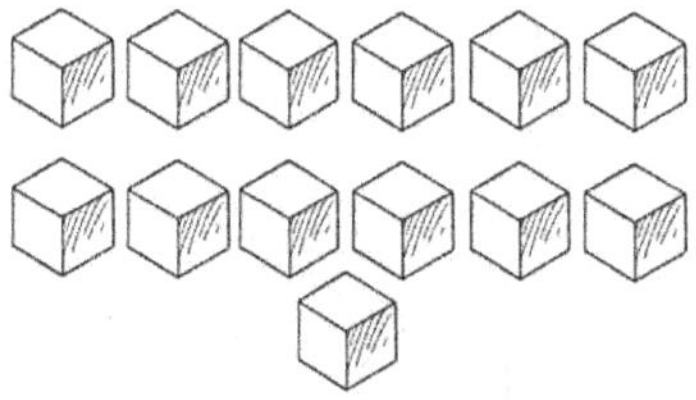

4.  11
   - 6
   ☐

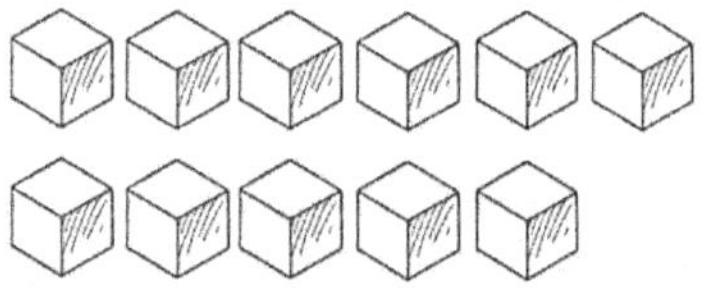

5.   6
   + 11
   ☐

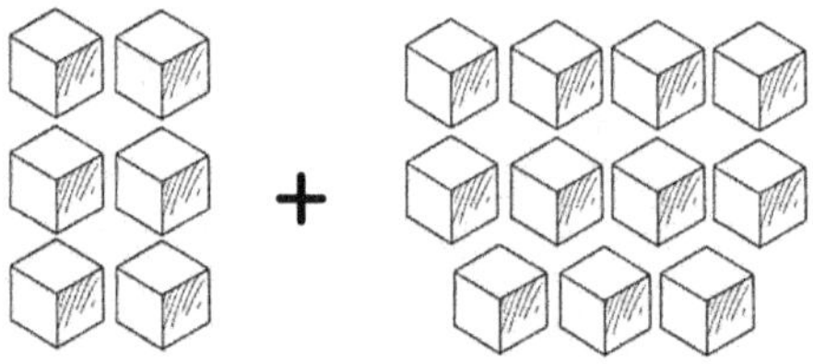

6.  14
   + 3
   ☐

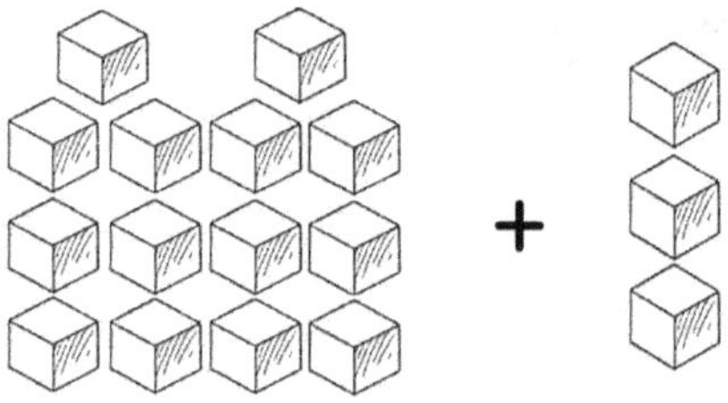

7.  16
   - 3
   ☐

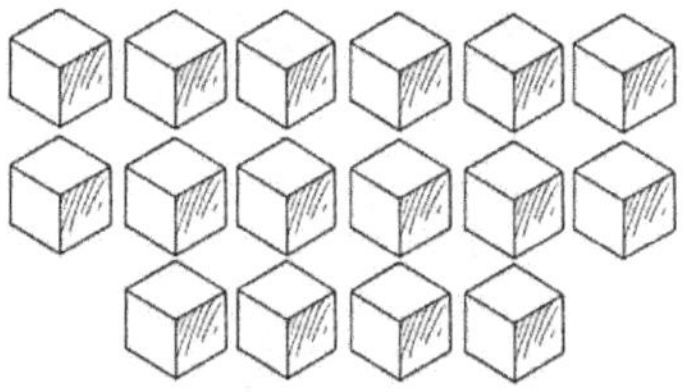

8.  17
   - 8
   ☐

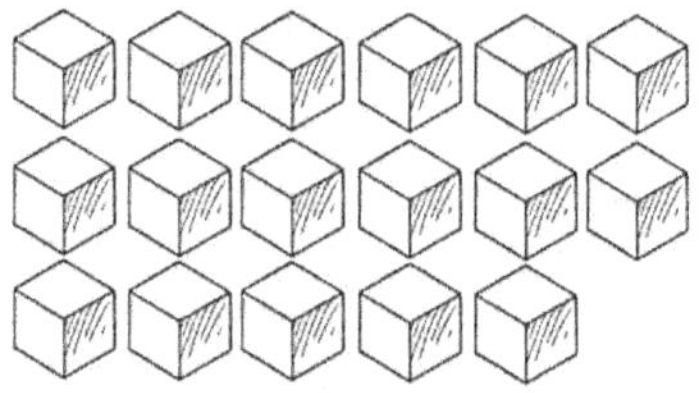

9.   4
   + 12
   ☐

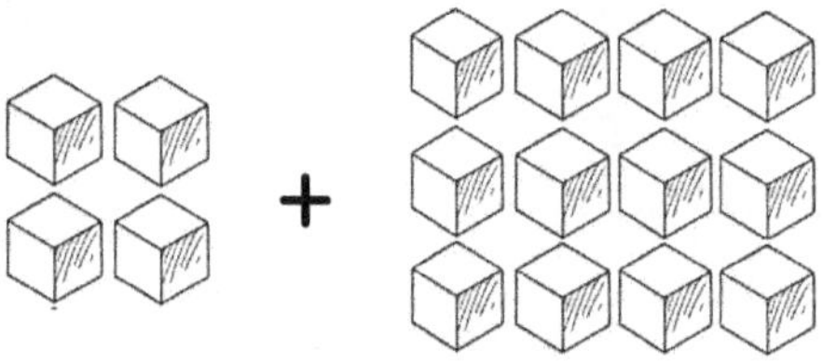

10. 15
   + 4
   ☐

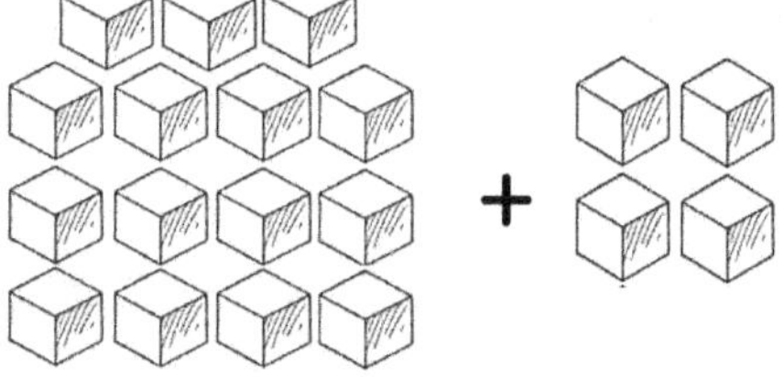

11. 19
   - 7
   ☐

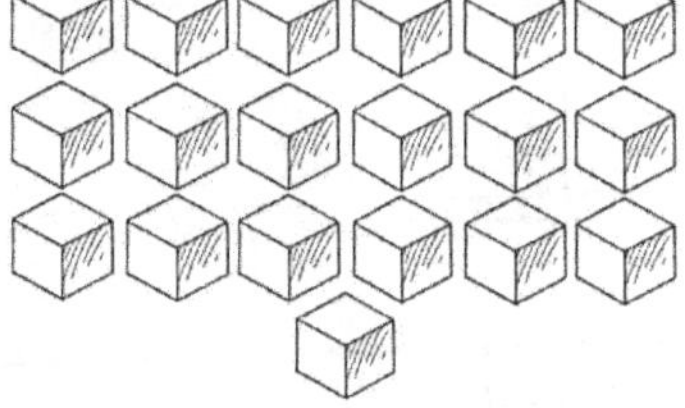

12. 18
   - 9
   ☐

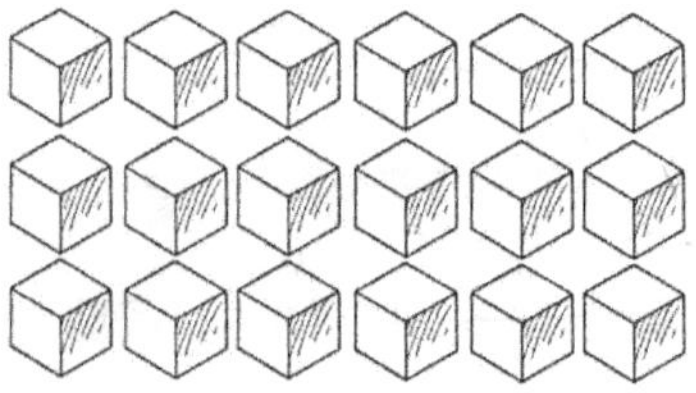

©2008 Plurarch Publications, Inc. PPI - 2001

Name _____    Standard: Facts to 20

Add or subtract the items then write the answer below each problem.

| | | | |
|---|---|---|---|
| 1.    10 <br> + 3 | 2.    11 <br> + 4 | 3.    14 <br> + 2 | 4.    12 <br> + 5 |
| 5.    10 <br> − 7 | 6.    13 <br> − 5 | 7.    15 <br> − 7 | 8.    11 <br> − 4 |
| 9.    9 <br> + 8 | 10.    11 <br> + 6 | 11.    13 <br> + 7 | 12.    12 <br> + 5 |
| 13.    17 <br> − 8 | 14.    16 <br> − 8 | 15.    14 <br> − 6 | 16.    18 <br> − 9 |
| 17.    15 <br> + 5 | 18.    12 <br> + 6 | 19.    7 <br> + 10 | 20.    13 <br> + 1 |
| 21.    12 <br> − 7 | 22.    15 <br> − 9 | 23.    19 <br> − 7 | 24.    20 <br> − 10 |

©2008 Plurarch Publications, Inc. PPI - 2001

Name _____    Standard: Order Sets

Count the items in each pair of boxes. Write the number on the line. Circle the number that is less than the other (smaller). Circle both if they are equal (same number).

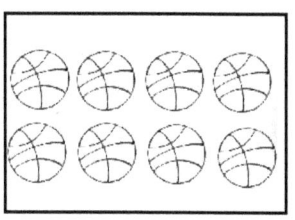 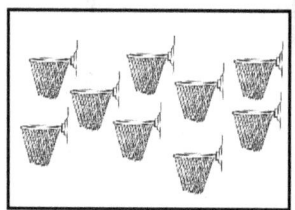  

_____  _____    _____  _____

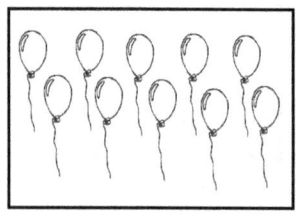 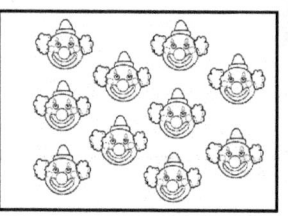 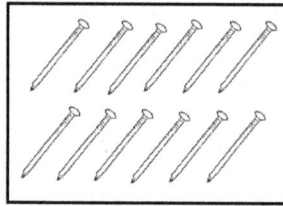 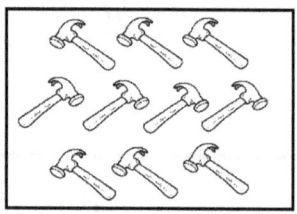

_____  _____    _____  _____

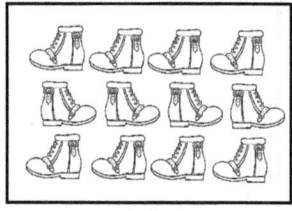 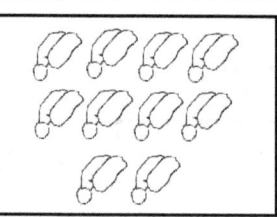 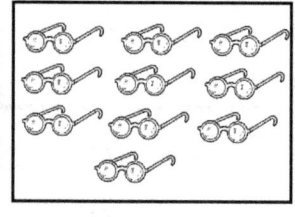 

_____  _____    _____  _____

 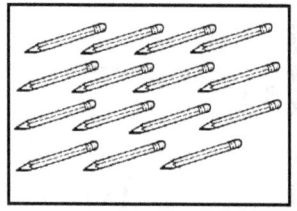 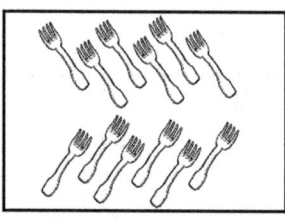 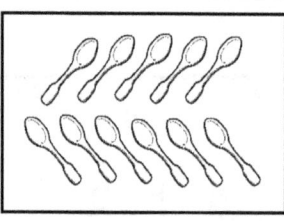

_____  _____    _____  _____

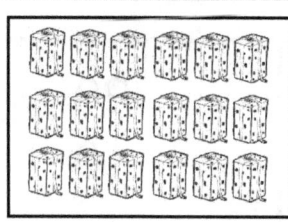   

_____  _____    _____  _____

©2008 Plurarch Publications, Inc. PPI - 2001

Name _____    Standard: Order Sets

Write the number under each box. Put a "**M**" on the box that has the most, an "**X**" on the box that has least, and circle the two boxes that are equal.

1.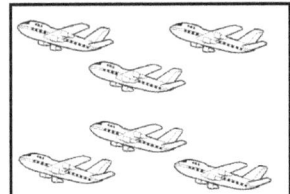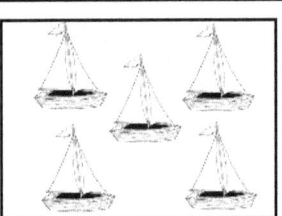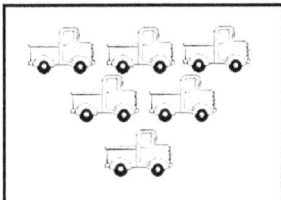

   _____    _____    _____    _____

2.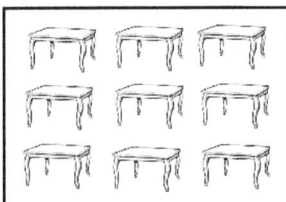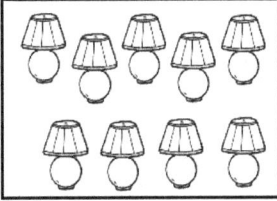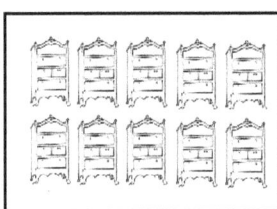

   _____    _____    _____    _____

3.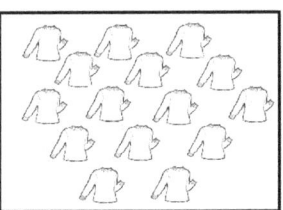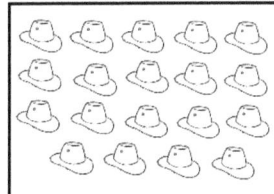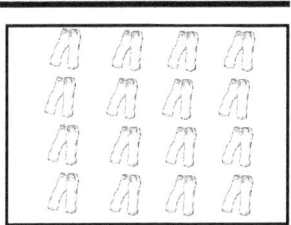

   _____    _____    _____    _____

4.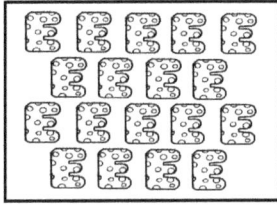

   _____    _____    _____    _____

5.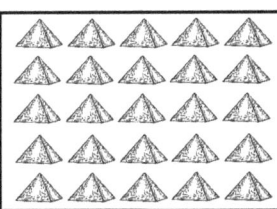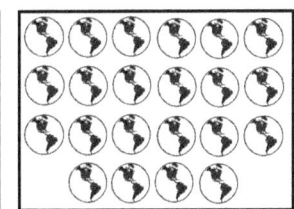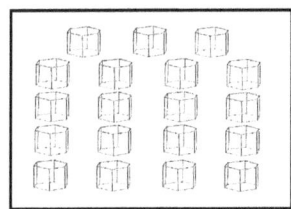

   _____    _____    _____    _____

©2008 Plurarch Publications, Inc. PPI - 2001

Name _____  Standard: Order Sets

Answer the problems in each box. Circle the answer that is greater than the other (more). Circle both numbers if they are equal (same number).

| | |
|---|---|
| 9      13 <br> +2     -1 | 10     12 <br> +2     -2 |
| 8      9 <br> +8     +7 | 12     4 <br> -6     +4 |
| 13     7 <br> -2     +5 | 10     8 <br> +6     +9 |
| 14     11 <br> -7     -8 | 10     18 <br> +4     -4 |
| 20     14 <br> -5     +2 | 17     20 <br> +3     -2 |

©2008 Plurarch Publications, Inc. PPI - 2001

Name _____     Standard: Word Problems

Read the story then answer the questions below.

Pam  has six red fish.

Sam  has eight blue fish.

Tam  has two yellow fish.

| | |
|---|---|
| 1. How many fish are blue? | 2. How many fish does Tam have? |
| 3. Who has the least number of fish?<br><br>Pam?   Sam?   Tam?<br>(circle one) | 4. Who has the largest number of fish?<br><br>Pam?   Sam?   Tam?<br>(circle one) |
| 5. How many fish do Pam and Tam have all together? | 6. How many fish do Sam and Pam have all together? |
| 7. How many fish do Tam and Sam have all together? | 8. If Sam bought two more blue fish, how many would he have all together? |
| 9. Pam bought three more red fish. Now how many does she have all together? | 10. Tam's Mother gave her seven more yellow fish. Now how many does she have? |

©2008 Plurarch Publications, Inc. PPI - 2001

Name _____    Standard: Word Problems

Read the story then answer the questions below.

Andy has seven black dogs.
Randy has nine tan dogs.
Sandy has four brown dogs.

| 1. How many dogs are black? | 2. Are the greatest number of dogs: <br><br> Black    Brown    Tan <br><br> (circle one) |
|---|---|
| 3. Who has more dogs: <br><br> Andy     Randy  <br><br> (circle one) | 4. Who has less dogs: <br><br> Andy     Sandy  <br><br> (circle one) |
| 5. Andy has how many more dogs than Sandy? | 6. Randy has how many more dogs than Andy? |
| 7. Randy has how many more dogs than Sandy? | 8. Randy gave away four of his dogs. How many dogs does he have left? |
| 9. Six of Randy's dogs went to sleep. How many of his dogs are still awake? | 10. Four of the tan dogs are eating. How many tan dogs are not eating? |

©2008 Plurarch Publications, Inc. PPI - 2001

Name _____    Standard: Word Problems

Read the story then answer the questions below.

---

Six ducks  were swimming in the pond.

Three frogs  were swimming in the pond.

Four ducks  flew away, then one frog  hopped away.

---

| | |
|---|---|
| 1. How many ducks were swimming in the pond at first? | 2. How many frogs were swimming in the pond at first? |
| 3. How many ducks flew away? | 4. How many frogs hopped away? |
| 5. At first, how many ducks and frogs were swimming in the pond all together? | 6. How many ducks were left after four flew away? |
| 7. How many frogs were left after one hopped away? | 8. At first, how many more ducks than frogs were swimming in the pond? |
| 9. How many frogs and ducks left the pond all together? | 10. How many more ducks flew away than frogs hopped away? |

Name _____    Standard: Word Problems

Read the story then answer the questions below.

Fran  has two cookies.    Stan  has three cookies.

Dan  has five cookies.    Nan  has eight cookies.

| | |
|---|---|
| 1. Who has more cookies:<br><br>Fran     Stan <br>(circle one) | 2. Who has less cookies:<br><br>Dan     Nan <br>(circle one) |
| 3. Who has the most cookies:<br><br>Dan   Fran   Nan <br>(circle one) | 4. Who has the least cookies:<br><br>Dan   Stan   Fran<br>(circle one) |
| 5. Nan has how many more cookies than Fran? | 6. How many cookies do Dan and Fran have together? |
| 7. How many cookies do Nan and Fran have together? | 8. Dan has how many more cookies than Stan? |
| 9. How many cookies do Stan and Dan have together? | 10. Nan has how many more cookies than Dan? |

©2008 Plurarch Publications, Inc. PPI - 2001

Name _____   Standard: Word Problems

Read the story then answer the questions below.

Ten squirrels  were sitting in a tree.  Six of the squirrels  left to find some nuts.  Two of the squirrels  jumped to another tree.  The squirrels found nine nuts.  They had a good lunch.

| | |
|---|---|
| 1. In the beginning, how many squirrels were sitting in the tree? | 2. How many squirrels first went to look for nuts? |
| 3. How many nuts did the squirrels find? | 4. How many squirrels jumped to another tree? |
| 5. How many squirrels were left in the tree when six left to find nuts? | 6. How many squirrels were still in the tree when two jumped to another tree? |
| 7. How many squirrels left the tree all together? | 8. How many more squirrels went to look for nuts than jumped to another tree? |
| 9. What did the squirrels do with the nuts? | 10. There were how many more squirrels than nuts? |

©2008 Plurarch Publications, Inc. PPI - 2001

Name _____  Standard: Word Problems

Read the story then answer the questions below.

**Cara has a lot of books in her bedroom. She has six books about cats, three books about dogs, seven books about birds, and eleven books about cooking.**

| | |
|---|---|
| 1. What is the name of the girl in this problem? | 2. How many books about cats does Cara have? |
| 3. Cara has how many books about dogs? | 4. How many books does she have that are about birds? |
| 5. She has how many books about cooking? | 6. How many books about cats and dogs together? |
| 7. How many of Cara's books are about animals? | 8. Cara has how many more books about birds than dogs? |
| 9. How many more books are about cooking than birds? | 10. How many books does Cara have in all? |
| 11. Where does Cara keep her books? | 12. Which type of book is Cara's favorite? |

©2008 Plurarch Publications, Inc. PPI - 2001

Name _____     Standard: Word Problems

Read the story then answer the questions below.

> **Liam went to the store for some candy. He walked three blocks west and two blocks north. He bought a lollipop for ten cents, a bag of chips for twenty-five cents, and a pack of gum for fifteen cents.**

| | |
|---|---|
| 1. What is the name of the boy in this problem? | 2. How many blocks did Liam walk in all? |
| 3. How many blocks to the south did Liam walk? | 4. What three things did Liam buy at the store? |
| 5. How much did the bag of chips cost? | 6. How much did the chips and lollipop cost together? |
| 7. Liam walked how many more blocks west than north? | 8. What was the name of the store? |
| 9. How much more did the chips cost than the lollipop? | 10. Which item cost Liam the least amount of money? |
| 11. How much money did Liam spend for these three items? | 12. Which item cost Liam the most money? |

©2008 Plurarch Publications, Inc. PPI - 2001

Name _____    Standard: Word Problems

Read the story then answer the questions below.

> **It rained four inches on Monday, two inches on Tuesday, six inches on Wednesday, and one inch on Thursday. Friday and Saturday the sun was shining and it did not rain a drop!**

| | |
|---|---|
| 1. How much rain fell on Thursday? | 2. On which day did it rain the most? |
| 3. How much more rain fell Wednesday than Monday? | 4. How much rain fell on Monday and Friday together? |
| 5. How much rain fell all together? | 6. On which days didn't it rain at all? |
| 7. What was the weather like on Sunday? | 8. How much rain fell on Monday, Tuesday and Thursday? |
| 9. Which two days did it rain most?<br><br>   Monday      Wednesday<br>    and           and<br>  Tuesday     Thursday<br>      (circle one) | 10. How much rain fell on the days starting with the letter "T"? |
| 11. How many days did it rain in all? | 12. On which rainy day did the least amount of rain fall? |

©2008 Plurarch Publications, Inc. PPI - 2001

Name _____   Standard: Word Problems

Read the story then answer the questions below.

> **Sloan had one quarter, two dimes, one nickel, and seven pennies. He bought two pencils for sixteen cents. He paid eight more cents to get a sticker. Now Sloan wants to buy a book that costs twenty-nine cents.**

| | |
|---|---|
| 1. How many pennies did Sloan have at first? | 2. Where did Sloan buy these items? |
| 3. How many pencils did Sloan buy? | 4. How much money did Sloan have at first? |
| 5. How much did Sloan have left after he bought the pencils? | 6. How much money did Sloan spend on pencils and stickers? |
| 7. How much does the book cost? | 8. Does Sloan have enough money left to buy the book? |
| 9. What is the book about? | 10. Which is worth more?  two dimes and seven pennies    one quarter  (circle one) |
| 11. Which cost more, the pencils or the sticker? | 12. How much did the sticker cost Sloan? |

©2008 Plurarch Publications, Inc. PPI - 2001

Name _____   Standard: Count By Twos

**Directions:** Circle each set of two items.  Fill in the blanks with the number of items.

1. (1  2)  3  4  5  6  7  8  9  10
   **1  2  3  4  5**  ___  ___  ___  ___  ___

2.  1  2  3  4  5  6  7  8  9  10
        **2**     **4**     **6**        ___        ___

3.  1  2  3  4  5  6  7  8  9  10
        **2**     **4**     ___     ___     ___

4.  A  A  A  A  A  A  A  A  A  A
      ___      ___      ___      ___      ___

5.  Z  Z  Z  Z  Z  Z  Z  Z  Z  Z
      ___      ___      ___      ___      ___

6.
      ___      ___      ___      ___      ___

7.
      ___      ___      ___      ___      ___

8.
      ___      ___      ___      ___      ___

©2008 Plurarch Publications, Inc. PPI - 2001

Name _____   Standard: Count By Twos

**Directions:** Circle each set of two items. Fill in the blanks with the number of items.

1. (1   2)   3   4   5   6   7   8   9   10
    1   2   3   4   5   ___  ___  ___  ___  ___

2. R   R   R   R   R   R   R   R   R   R
        2       4       6       ___

3. M   M   M   M   M   M   M   M   M   M
        2       4       ___     ___     ___

4. ✖ ✖ ✖ ✖ ✖ ✖ ✖ ✖ ✖ ✖
  ___   ___   ___   ___   ___

5. ★ ★ ★ ★ ★ ★ ★ ★ ★ ★
  ___   ___   ___   ___   ___

6.
  ___   ___   ___   ___   ___

7.
  ___   ___   ___   ___   ___

8.
  ___   ___   ___   ___   ___

©2008 Plurarch Publications, Inc. PPI - 2001

Name _____  Standard: Count By Twos

**Directions:** Circle each set of two items. Fill in the blanks with the number of items.

1.

   ___    ___    ___    ___    ___

2.

   ___    ___    ___    ___    ___

3.   1   2   3   4   5   6   7   8   9   10

   ___    ___    ___    ___    ___

   11   12   13   14   15   16   17   18   19   20

   ___    ___    ___    ___    ___

   21   22   23   24   25   26   27   28   29   30

   ___    ___    ___    ___    ___

4.

   ___    ___    ___    ___    ___

   ___    ___    ___    ___    ___

   ___    ___    ___    ___    ___

©2008 Plurarch Publications, Inc. PPI - 2001

Name _____   ( Standard: Count By Twos )

**Directions:** Count by twos. Fill in the missing numbers

1. | 2 | 4 | 6 |   | 10 | 12 | 14 |   | 18 |   |

2. | 10 |   | 14 |   | 18 |   | 22 |   | 26 |   |

3. |   | 20 |   | 24 |   | 28 |   | 32 |   | 36 |

4. | 26 |   |   | 32 |   |   | 38 |   |   | 44 |

5. |   | 42 |   |   | 48 |   |   | 54 |   |   |

6. | 52 |   |   |   | 60 |   |   |   | 68 |   |

7. | 56 |   |   |   | 64 |   |   |   | 72 |   |

8. |   |   | 68 |   |   |   |   | 78 |   |   |

9. |   | 78 |   |   |   |   | 88 |   |   |   |

10. | 82 |   |   |   |   |   |   |   |   |   |

Name _____  Standard: Count By Fives

**Directions:** Circle each set of five items.  Fill in the blanks with the number of items.

1. ( 1   2   3   4   5 )   6   7   8   9   10
   __1__  __2__  __3__  __4__  __5__   ___   ___   ___   ___   ___

2.   1   2   3   4   5   6   7   8   9   10
                     __5__                           ___

3.   1   2   3   4   5   6   7   8   9   10
                     ___                             ___

4.   A   A   A   A   A   A   A   A   A   A
                     ___                             ___

5.   Z   Z   Z   Z   Z   Z   Z   Z   Z   Z
                     ___                             ___

6.   (mitten images ×10)
                     ___                             ___

7.   (sock images ×10)
                     ___                             ___

8.   (shoe images ×10)
                     ___                             ___

©2008 Plurarch Publications, Inc. PPI - 2001    48

Name _____   Standard: Count By Fives

**Directions:** Circle each set of five items. Fill in the blanks with the number of items.

1. ( 1   2   3   4   5 )   6   7   8   9   10

   __5__                                __ __

2.

   __5__                                __ __

3.

   __ __                                __ __

4. X   O   X   O   X   O   X   O   X   O

   __ __                                __ __

5. 1   2   3   4   5   6   7   8   9   10

   __ __                                __ __

   11   12   13   14   15   16   17   18   19   20

   __ __                                __ __

6. A   B   C   D   E   F   G   H   I   J

   __ __                                __ __

   K   L   M   N   O   P   Q   R   S   T

   __ __                                __ __

©2008 Plurarch Publications, Inc. PPI - 2001

Name _____   Standard: Count By Fives

**Directions:** Count by fives. Fill in the missing numbers

1. | 5 | 10 | 15 |   | 25 | 30 | 35 |   | 45 |   |

2. | 20 |   | 30 |   | 40 |   | 50 |   | 60 |   |

3. | 15 |   | 25 |   | 35 |   | 45 |   | 55 |   |

4. | 10 |   |   | 25 |   |   | 40 |   |   | 55 |

5. | 35 |   |   | 50 |   |   | 65 |   |   |   |

6. | 5 |   |   |   | 25 |   |   |   | 45 |   |

7. | 35 |   |   |   | 55 |   |   |   | 75 |   |

8. |   |   | 30 |   |   |   |   | 55 |   |   |

9. | 45 |   |   |   |   |   |   | 75 |   |   |

10. | 55 |   |   |   |   |   |   |   |   |   |

©2008 Plurarch Publications, Inc. PPI - 2001

Name _____    Standard: Count By Tens

**Directions:** Circle each set of ten items. Fill in the blanks with the number of items.

1. ( 1   2   3   4   5   6   7   8   9   10 )
   <u>1</u>  <u>2</u>  <u>3</u>  <u>4</u>  <u>5</u>   __   __   __   __   __

2.   1   2   3   4   5   6   7   8   9   10
   __

3.   H   H   H   H   H   H   H   H   H   H
   __

4.   (mittens ×10)
   __

5.   (socks ×10)
   __

6.   (shoes ×10)
   __

7.   A   A   A   A   A   A   A   A   A   A
   __

8.   Z   Z   Z   Z   Z   Z   Z   Z   Z   Z
   __

©2008 Plurarch Publications, Inc. PPI - 2001

Name _____   Standard: Count By Tens

**Directions:** Circle each set of ten items. Fill in the blanks with the number of items.

1.

    ____

2.

    ____

3. 1   2   3   4   5   6   7   8   9   10

    ____

    11   12   13   14   15   16   17   18   19   20

    ____

    21   22   23   24   25   26   27   28   29   30

    ____

4.

    ____

    ____

    ____

©2008 Plurarch Publications, Inc. PPI - 2001

Name _____  Standard: Count By Tens

The library is putting away old books. The books are stacked in groups of ten. Count by tens to find how many books are in each of the problem boxes below.

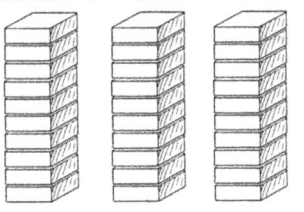  30

10 + 10 + 10 = 30 books

1. How many books?
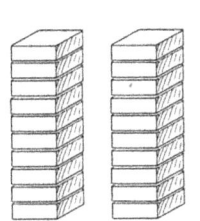

2. How many books?
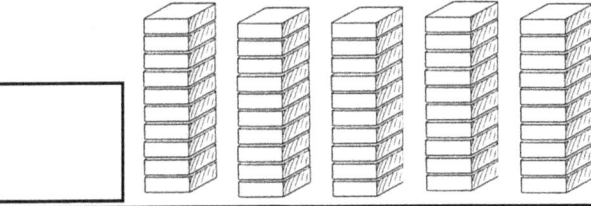

3. How many books?
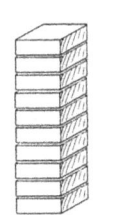

4. How many books?
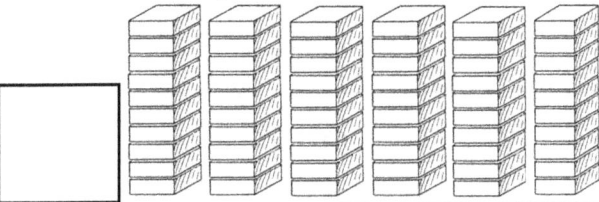

5. How many books?
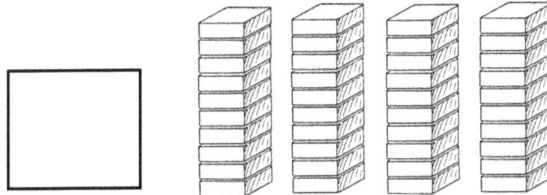

6. How many books?
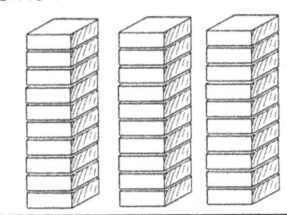

7. How many books?
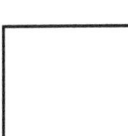 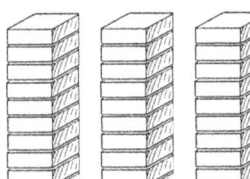  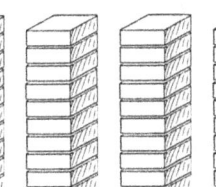 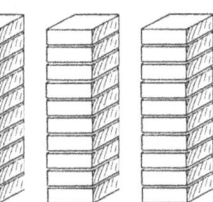

8. How many books?
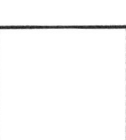 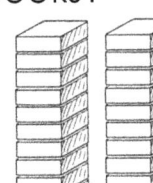 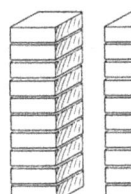 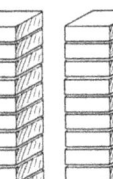 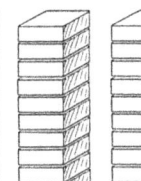 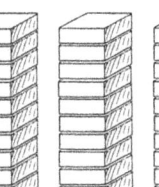 

©2008 Plurarch Publications, Inc. PPI - 2001

Name _____  Standard: Time Measurement

Read the time words and their meanings, then answer the questions.

**day** - 24 hours from midnight until the next midnight
**daylight** - the time from sunrise until sunset
**morning** - the time from sunrise until 12 o'clock noon
**noon** - 12 o'clock in the middle of the day
**afternoon** - the time from noon until 6 o'clock.
**evening** - near the end of the day, around sunset
**night** - the time when it is dark outside
**midnight** - 12 o'clock at night, the middle of the night

1. What word tells us the sun is out?

   daylight          night          midnight

2. What word tells us it is dark outside?

   morning          afternoon          night

3. Which word means the middle of the night?

   daylight          evening          midnight

4. When do we eat lunch?

   morning          noon          evening

5. When do we eat dinner?

   morning          noon          evening

6. When do most people sleep?

   morning          noon          night

©2008 Plurarch Publications, Inc. PPI - 2001

Name _____   Standard: Time Measurement

Here are two ways to measure time. Use them to answer the questions below.

| clock | calendar |
|---|---|
| hours   minutes | days   months |

1. Which shows the time of day?

   clock      calendar

2. Which shows the days of the week?

   clock      calendar

3. Which tells the months of the year?

   clock      calendar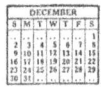

4. Which tells minutes and hours?

   clock      calendar

5. Which would you use to plan for next month?

   clock      calendar

6. Which would you use to plan dinner time?

   clock      calendar

©2008 Plurarch Publications, Inc. PPI - 2001

Name _____   Standard: Time to the Hour

Write the time shown on each clock.

| Clock | Time | Clock | Time |
|---|---|---|---|
| 5:00 | _____ | 3:00 | _____ |
| 8:00 | _____ | 11:00 | _____ |
| 12:00 | _____ | 10:00 | _____ |
| 9:00 | _____ | 6:00 | _____ |
| 7:00 | _____ | 1:00 | _____ |

Name _____    Standard: Time to the Half Hour

Write the time shown on each clock.

_____    _____

_____    _____

_____    _____

_____    _____

_____    _____

©2008 Plurarch Publications, Inc. PPI - 2001

Name _____   Standard: Time to the Half Hour

Draw the hands to show the time given for each clock.

| | | | |
|---|---|---|---|
|  | 7:00 _____ |  | 3:00 _____ |
|  | 2:30 _____ |  | 11:30 _____ |
|  | 4:00 _____ |  | 1:30 _____ |
|  | 8:30 _____ |  | 9:00 _____ |
|  | 6:00 _____ |  | 5:30 _____ |

Name _____    Standard: Money

A penny = 1¢     A nickel = 5¢     A dime = 10¢     A quarter = 25¢

Count the money in each box. Write the amount of money shown.

1. How much money?  ☐ ¢

2. How much money?  ☐ ¢

3. How much money?  ☐ ¢

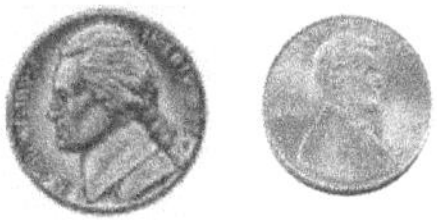

4. How much money?  ☐ ¢

5. How much money?  ☐ ¢

6. How much money?  ☐ ¢

7. How much money?  ☐ ¢

8. How much money?  ☐ ¢

©2008 Plurarch Publications, Inc. PPI - 2001

Name _____  Standard: Money

A penny = 1¢    A nickel = 5¢    A dime = 10¢    A quarter = 25¢

Count the money in each box. Write the amount of money shown.

1. How much money?  ☐ ¢

2. How much money?  ☐ ¢

3. How much money?  ☐ ¢

4. How much money?  ☐ ¢

5. How much money?  ☐ ¢

6. How much money?  ☐ ¢

7. How much money?  ☐ ¢

8. How much money?  ☐ ¢

©2008 Plurarch Publications, Inc. PPI - 2001

Name _____    Standard: Money

A penny = 1¢    A nickel = 5¢    A dime = 10¢    A quarter = 25¢

You can use different coins to make the same amount. For each row, show three different ways to make the amount shown.

| 12¢ = | | 12¢ = | | 12¢ = | |
|---|---|---|---|---|---|
| _____ | pennies | _____ | pennies | _____ | pennies |
| _____ | nickels | _____ | nickels | _____ | nickels |
| _____ | dimes | _____ | dimes | _____ | dimes |
| _____ | quarters | _____ | quarters | _____ | quarters |
| 14¢ = | | 14¢ = | | 14¢ = | |
| _____ | pennies | _____ | pennies | _____ | pennies |
| _____ | nickels | _____ | nickels | _____ | nickels |
| _____ | dimes | _____ | dimes | _____ | dimes |
| _____ | quarters | _____ | quarters | _____ | quarters |
| 19¢ = | | 19¢ = | | 19¢ = | |
| _____ | pennies | _____ | pennies | _____ | pennies |
| _____ | nickels | _____ | nickels | _____ | nickels |
| _____ | dimes | _____ | dimes | _____ | dimes |
| _____ | quarters | _____ | quarters | _____ | quarters |
| 23¢ = | | 23¢ = | | 23¢ = | |
| _____ | pennies | _____ | pennies | _____ | pennies |
| _____ | nickels | _____ | nickels | _____ | nickels |
| _____ | dimes | _____ | dimes | _____ | dimes |
| _____ | quarters | _____ | quarters | _____ | quarters |

©2008 Plurarch Publications, Inc. PPI - 2001

Name _____        Standard: Inches/Feet

Many years ago people used their hands to measure things. Hands come in many sizes, so the measurements were not always the same. A king decided that his foot should be the length used by everyone, so the **RULER** was invented! It is exactly one foot long. The foot was too big to measure smaller things so it was divided into 12 equal parts called **INCHES**. One inch is about the size of your thumb from the last knuckle to the tip.

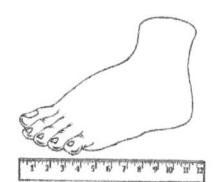

Find these things and measure their length. Measure with your hands first, then use a ruler. Circle if the item is bigger or smaller than one foot.

| 1. a pencil<br>_____ hands<br>_____ inches<br>BIGGER   SMALLER | 2. a desk or table top<br>_____ hands<br>_____ inches<br>BIGGER   SMALLER | 3. a science book<br>_____ hands<br>_____ inches<br>BIGGER   SMALLER |
|---|---|---|
| 4. your shoe bottom<br>_____ hands<br>_____ inches<br>BIGGER   SMALLER | 5. a window<br>_____ hands<br>_____ inches<br>BIGGER   SMALLER | 6. the seat of a chair<br>_____ hands<br>_____ inches<br>BIGGER   SMALLER |
| 7. a piece of paper<br>_____ hands<br>_____ inches<br>BIGGER   SMALLER | 8. a dictionary<br>_____ hands<br>_____ inches<br>BIGGER   SMALLER | 9. your own hand<br>_____ hands<br>_____ inches<br>BIGGER   SMALLER |
| 10. a computer screen<br>_____ hands<br>_____ inches<br>BIGGER   SMALLER | 11. a fork<br>_____ hands<br>_____ inches<br>BIGGER   SMALLER | 12. the waste basket<br>_____ hands<br>_____ inches<br>BIGGER   SMALLER |

©2008 Plurarch Publications, Inc. PPI - 2001

Name _____   Standard: Inches/Feet

Use **INCHES** to measure small things like buttons.
Use **FEET** to measure bigger things like walls.
Use **YARDS** to measure large things like buildings.
Use **MILES** to measure long distances.

12 inches = 1 foot
3 feet = 1 yard
5280 feet = 1 mile

Decide which measurement you would use for each item. Put an X next to the one you think is best.

| 1. a road | 2. a piece of paper | 3. the gym floor |
|---|---|---|
| ____ inches<br>____ feet<br>____ yards<br>____ miles | ____ inches<br>____ feet<br>____ yards<br>____ miles | ____ inches<br>____ feet<br>____ yards<br>____ miles |
| 4. a driveway | 5. a mailbox | 6. your teacher |
| ____ inches<br>____ feet<br>____ yards<br>____ miles | ____ inches<br>____ feet<br>____ yards<br>____ miles | ____ inches<br>____ feet<br>____ yards<br>____ miles |
| 7. a shoelace | 8. a necklace | 9. your back yard |
| ____ inches<br>____ feet<br>____ yards<br>____ miles | ____ inches<br>____ feet<br>____ yards<br>____ miles | ____ inches<br>____ feet<br>____ yards<br>____ miles |
| 10. a science book | 11. distance to a city | 12. length of a car |
| ____ inches<br>____ feet<br>____ yards<br>____ miles | ____ inches<br>____ feet<br>____ yards<br>____ miles | ____ inches<br>____ feet<br>____ yards<br>____ miles |

©2008 Plurarch Publications, Inc. PPI - 2001

Name _____  Standard: Fractions

**FRACTIONS** are equal sized parts of a whole. If you cut one apple into two equal parts and eat one part, you have eaten one out of two pieces, or 1/2 of the apple.

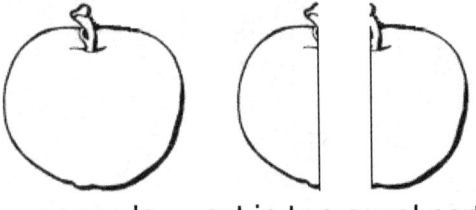

one apple    cut in two equal parts

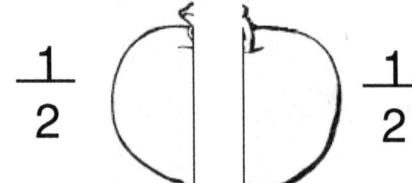

each part equals 1 out of two pieces or 1/2 of the apple

Cut each item below into two equal parts. Color 1/2 of each item.

Name _____   Standard: Fractions

**FRACTIONS** are equal sized parts of a whole. If you cut a sandwich into four equal parts and eat one part, you have eaten one out of four pieces, or 1/4 of the sandwich.

   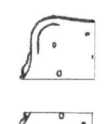 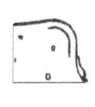   $\frac{1}{4}$   $\frac{1}{4}$

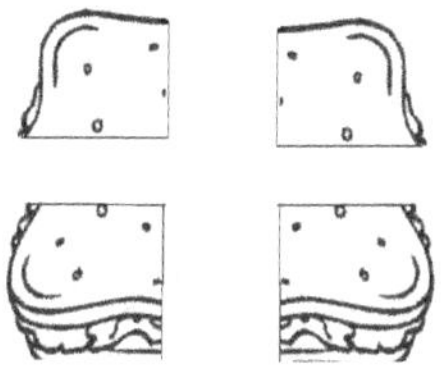

                         $\frac{1}{4}$  $\frac{1}{4}$

one sandwich    cut in four equal parts    each part equals 1 out of four pieces or 1/4 of the sandwich

Cut each item below into four equal parts. Color 1/4 of each item..

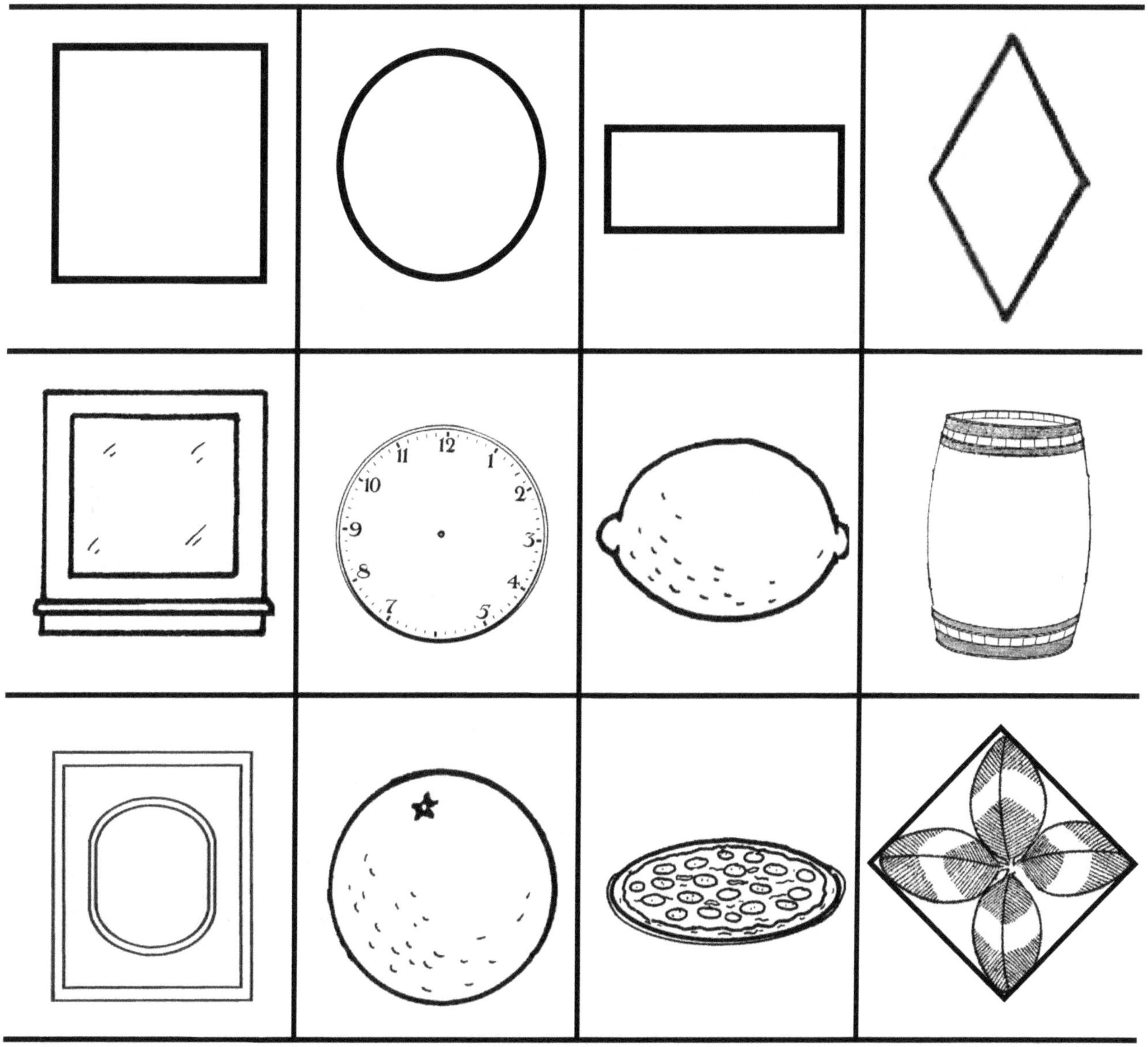

Name _____    Standard: Fractions

**FRACTIONS** are equal sized parts of a whole. The top number tells how many pieces are shaded. The bottom number tells how many pieces in all.

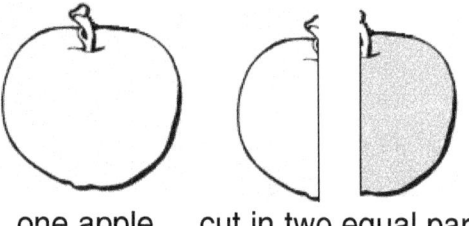

one apple    cut in two equal parts

$\frac{1}{2}$ 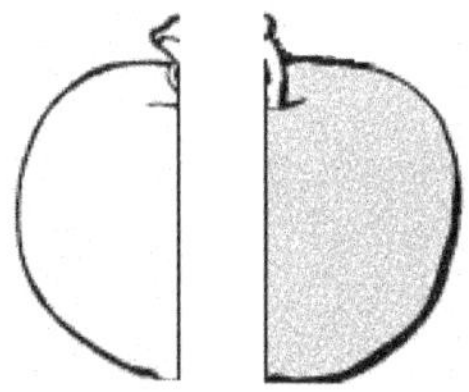   $\frac{1}{2}$

1 out of two pieces, or 1/2, is white
1 out of two pieces, or 1/2, is shaded

Tell what fraction is shaded for each picture below.

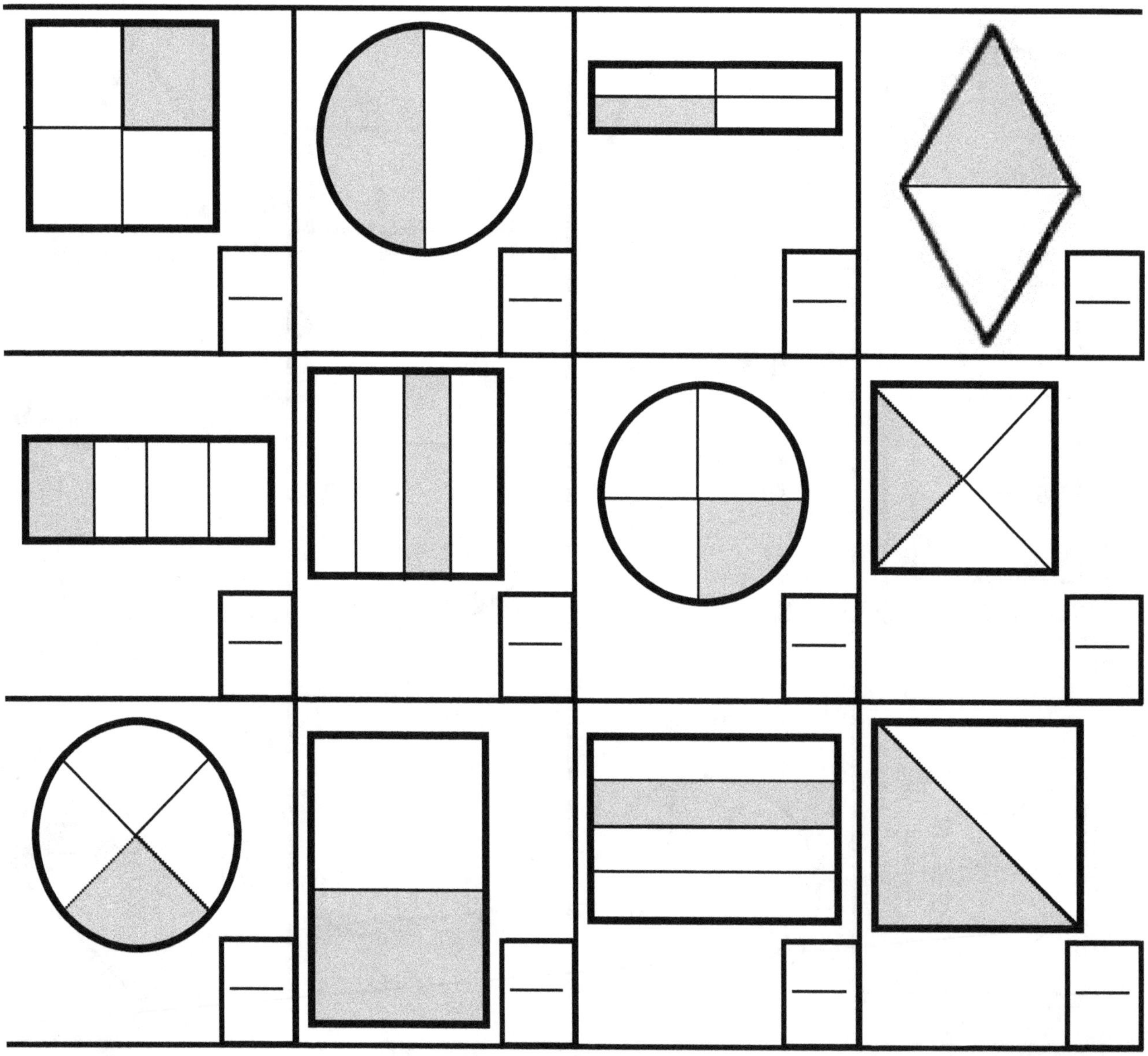

Name _____  Standard: Geometry

Look at the first item in each row. Put an **X** on any items in that row that have about the same shape as the first one.

Name _____   Standard: Geometry

Read the shape word.  Put an **X** on any items in that row that have that shape.

| CIRCLE | | | | |
|---|---|---|---|---|
| SQUARE | | | | |
| RECTANGLE | | | | |
| TRIANGLE | | | | |
| DIAMOND | | | | |
| CUBE | | | | |
| STAR | | | | |

©2008 Plurarch Publications, Inc. PPI - 2001

Name _____    Standard: Geometry

Draw the shape shown in each box. Can you name each shape?

# ANSWER KEYS: 6, 7, 8, 9

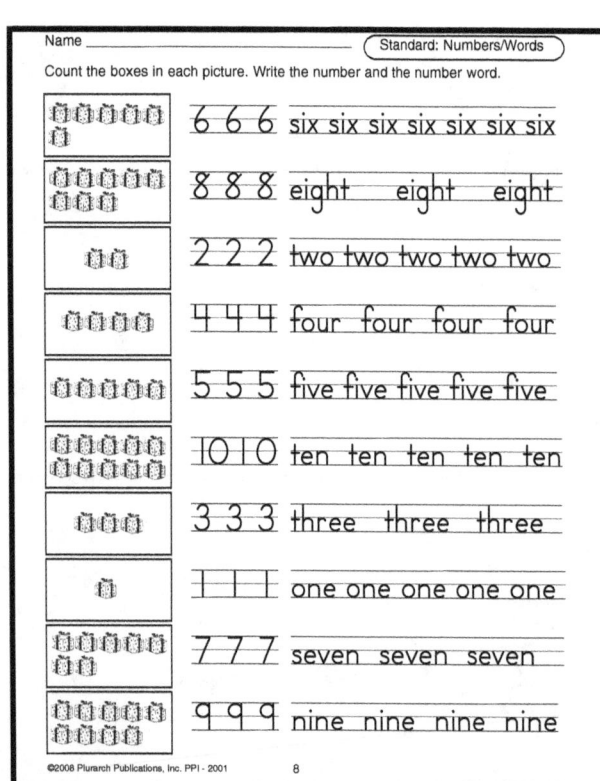

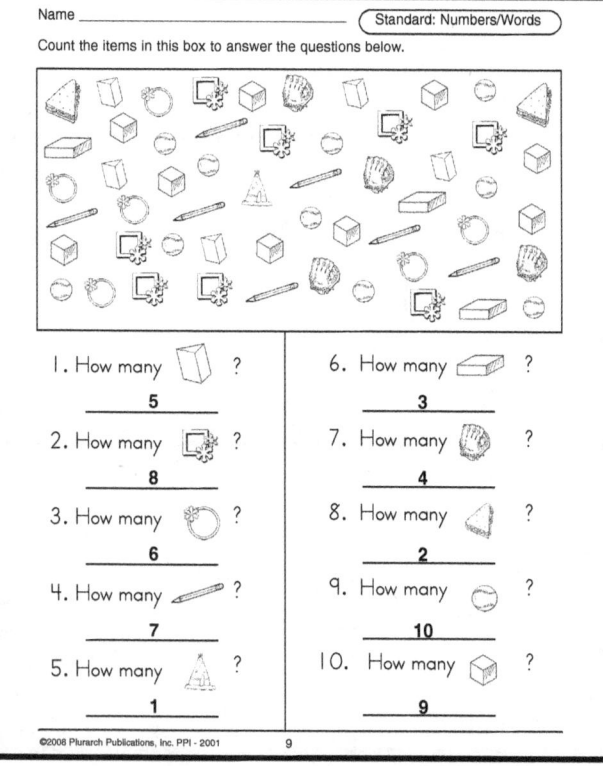

ANSWER KEYS: 10, 11, 12, 13

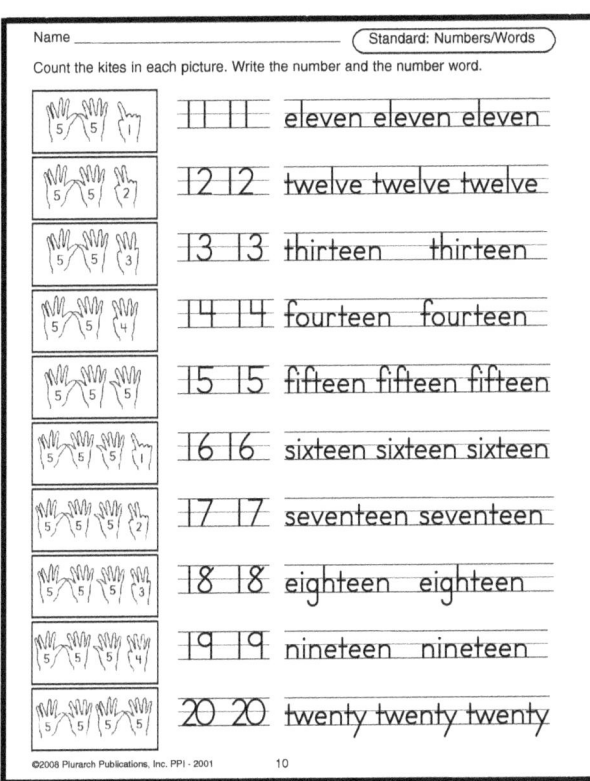

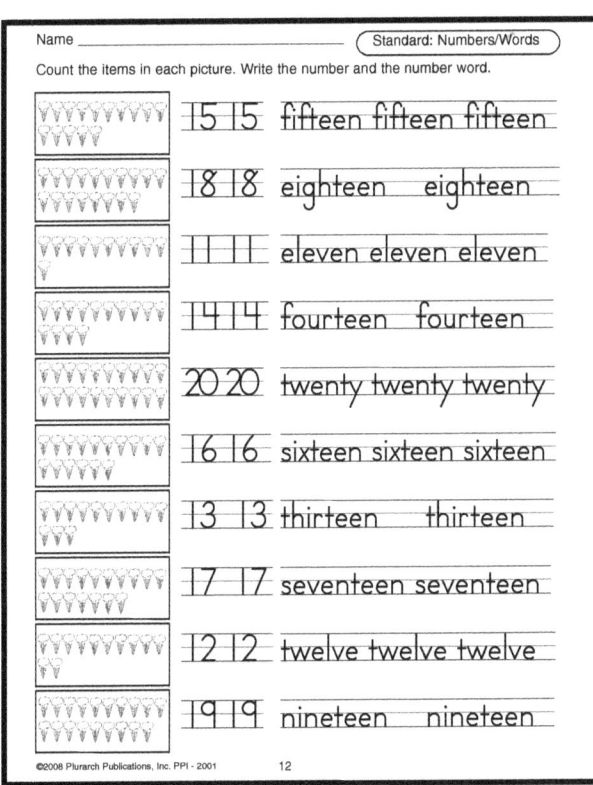

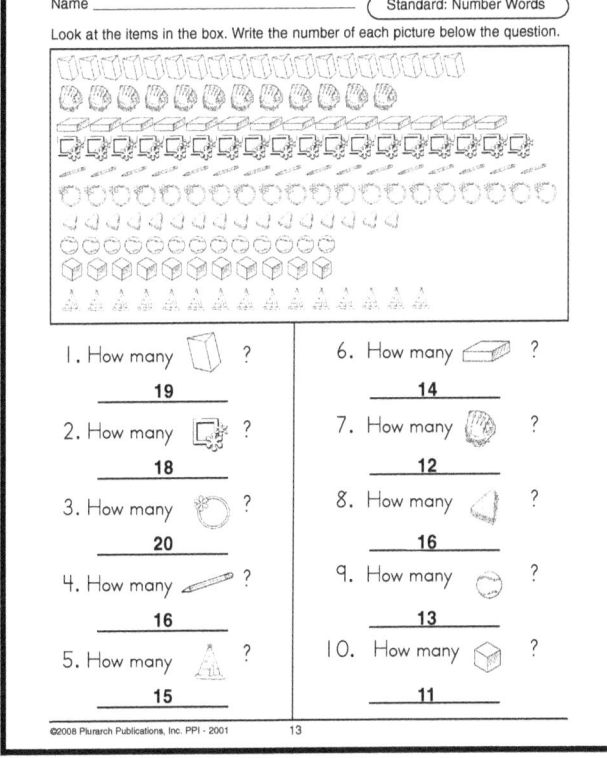

# ANSWER KEYS: 14, 15, 16, 17

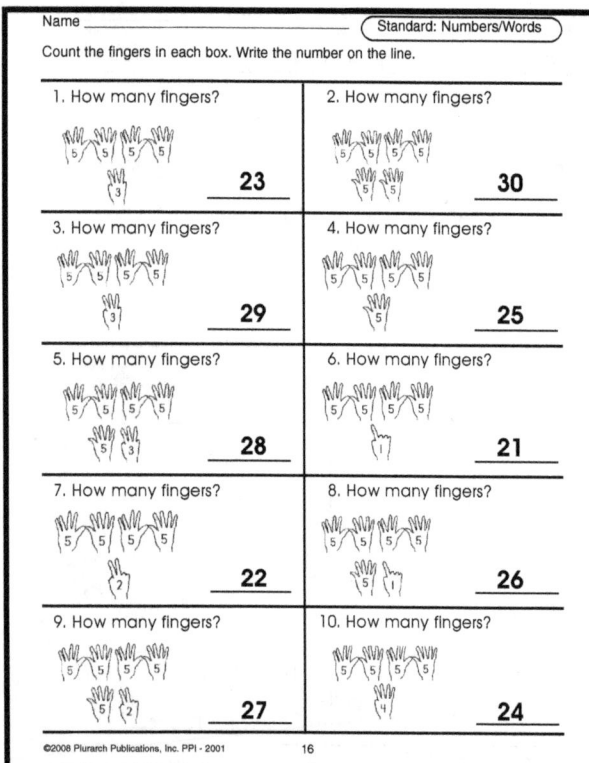

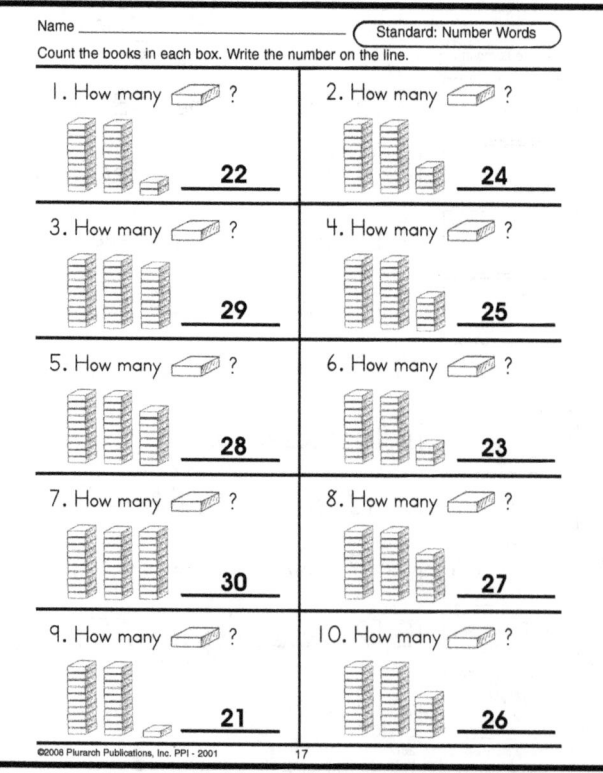

**ANSWER KEYS: 18, 19, 20, 21**

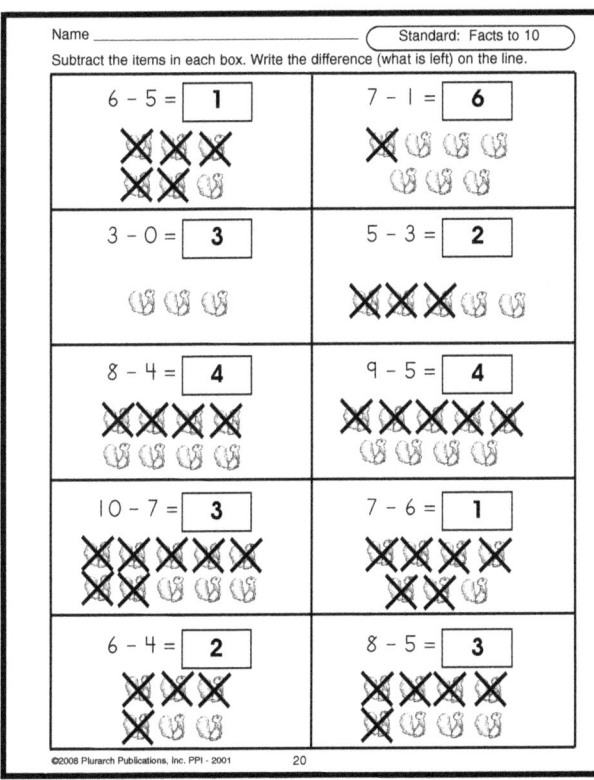
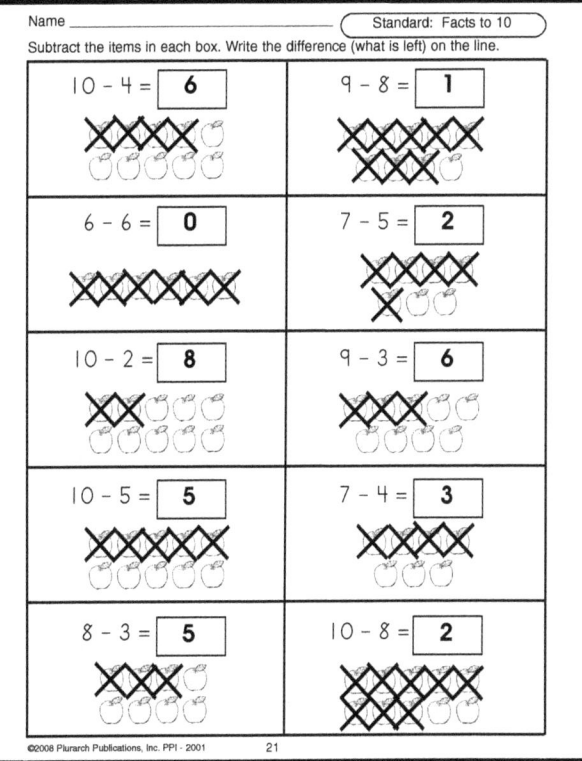

# ANSWER KEYS: 22, 23, 24, 25

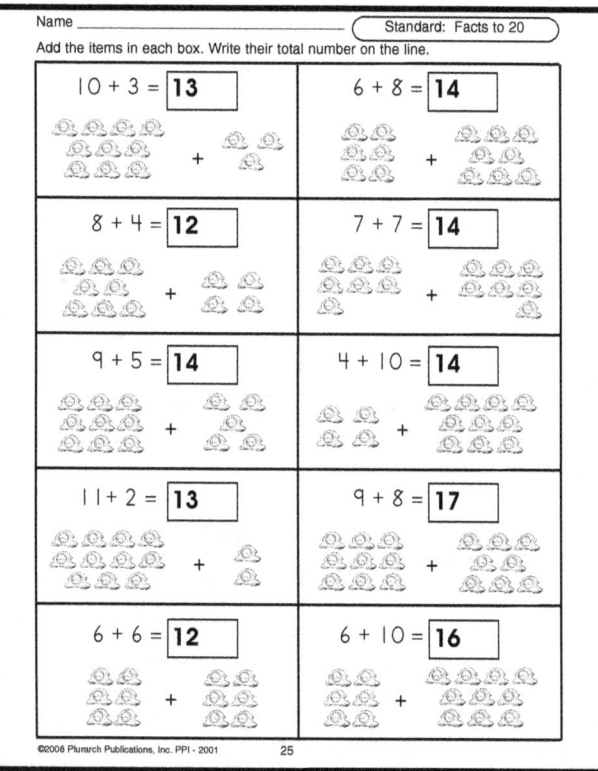

**ANSWER KEYS: 26, 27, 28, 29**

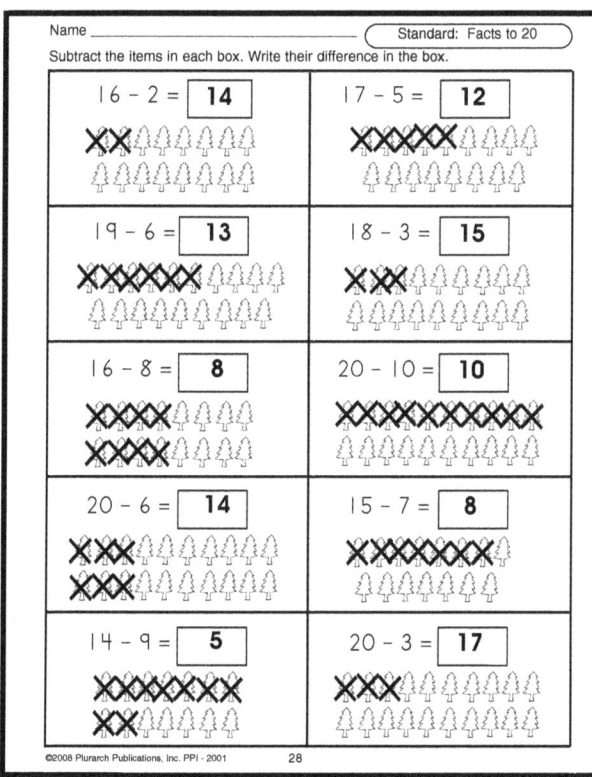

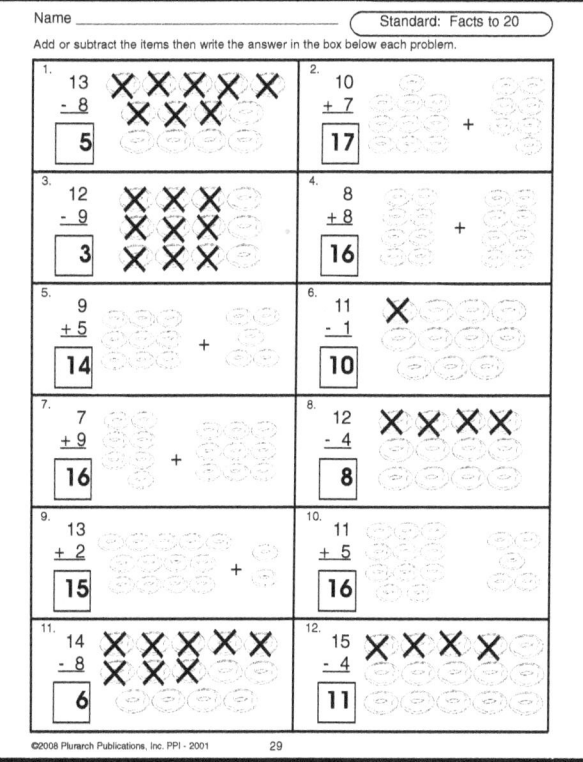

# ANSWER KEYS: 30, 31, 32, 33

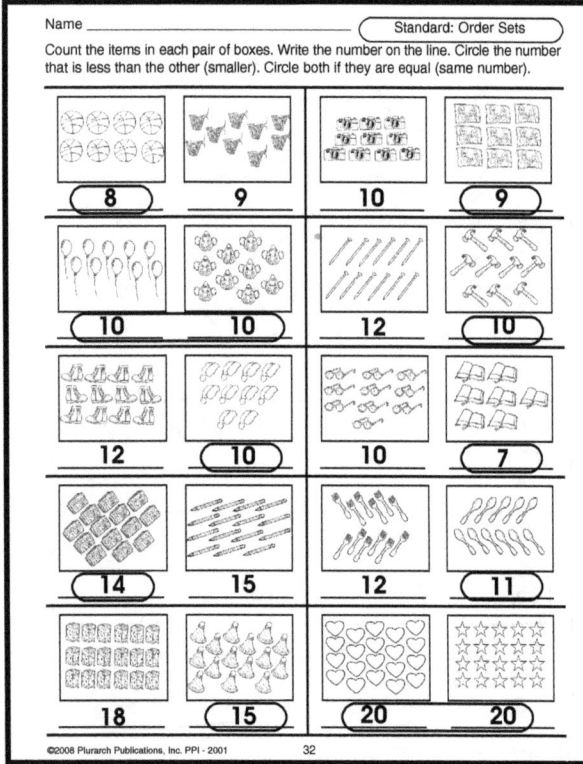

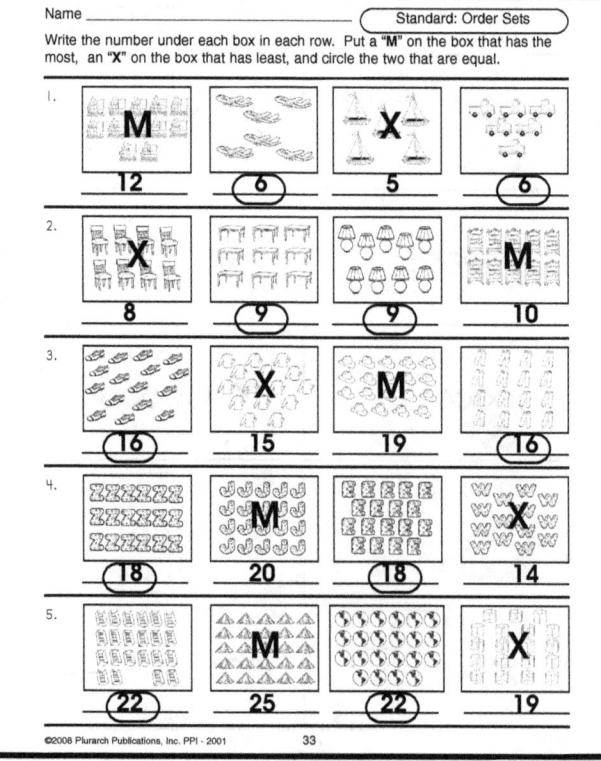

# ANSWER KEYS: 34, 35, 36, 37

## Page 34 — Standard: Order Sets

Answer the problems in each pair of boxes. Circle the answer that is greater than the other (more). Circle both numbers if they are equal (same number).

| | | | |
|---|---|---|---|
| 9 + 2 = 11 | 13 − 1 = (12) | 10 + 2 = (12) | 12 − 2 = 10 |
| 8 + 8 = (16) | 9 + 7 = (16) | 12 − 6 = 5 | 4 + 4 = (8) |
| 13 − 2 = 11 | 7 + 5 = (12) | 10 + 6 = 16 | 8 + 9 = (17) |
| 14 − 7 = (7) | 11 − 8 = 3 | 10 + 4 = (14) | 18 − 4 = (14) |
| 20 − 5 = 15 | 14 + 2 = (16) | 17 + 3 = (20) | 20 − 2 = 18 |

## Page 35 — Standard: Word Problems

Read the story then answer the questions below.

Pam has six red fish.
Sam has eight blue fish.
Tam has two yellow fish.

| | |
|---|---|
| 1. How many fish are blue? **8** | 2. How many fish does Tam have? **2** |
| 3. Who has the least number of fish: Pam? Sam? (Tam?) | 4. Who has the largest number of fish: Pam? (Sam?) Tam? |
| 5. How many fish do Pam and Tam have all together? **6 + 2 = 8** | 6. How many fish do Sam and Pam have all together? **8 + 6 = 14** |
| 7. How many fish do Tam and Sam have all together? **2 + 8 = 10** | 8. If Sam bought two more blue fish, how many would he have all together? **8 + 2 = 10** |
| 9. Pam bought three more red fish. Now how many does she have all together? **6 + 3 = 9** | 10. Tam's Mother gave her seven more yellow fish. Now how many does she have? **2 + 7 = 9** |

## Page 36 — Standard: Word Problems

Read the story then answer the questions below.

Andy has seven black dogs.
Randy has nine tan dogs.
Sandy has four brown dogs.

| | |
|---|---|
| 1. How many dogs are black? **7 black dogs** | 2. The greatest number of dogs are: Black   Brown   (Tan) |
| 3. Who has more dogs: Andy?  (Randy?) | 4. Who has less dogs: Andy?  (Sandy?) |
| 5. Andy has how many more dogs than Sandy? **7 − 4 = 3** | 6. Randy has how many more dogs than Andy? **9 − 7 = 2** |
| 7. Randy has how many more dogs than Sandy? **9 − 4 = 5** | 8. Randy gave away four of his dogs. How many dogs does he have left? **9 − 4 = 5** |
| 9. Six of Randy's dogs went to sleep. How many of his dogs are still awake? **9 − 6 = 3** | 10. Four of the tan dogs are eating. How many tan dogs are not eating? **9 − 4 = 5** |

## Page 37 — Standard: Word Problems

Read the story then answer the questions below.

Six ducks were swimming in the pond.
Three frogs were swimming in the pond.
Four ducks flew away, then one frog hopped away.

| | |
|---|---|
| 1. How many ducks were swimming in the pond at first? **6** | 2. How many frogs were swimming in the pond at first? **3** |
| 3. How many ducks flew away? **4** | 4. How many frogs hopped away? **1** |
| 5. At first, how many ducks and frogs were swimming in the pond all together? **6 + 3 = 9** | 6. How many ducks were left after four flew away? **6 − 4 = 2** |
| 7. How many frogs were left after one hopped away? **3 − 1 = 2** | 8. At first, how many more ducks than frogs were swimming in the pond? **6 − 3 = 3** |
| 9. How many frogs and ducks left the pond all together? **4 + 1 = 5** | 10. How many more ducks flew away than frogs hopped away? **4 − 1 = 3** |

**ANSWER KEYS: 38, 39, 40, 41**

---

Name _____   Standard: Word Problems
Read the story then answer the questions below.

Fran had two cookies.   Stan had three cookies.
Dan had five cookies.   Nan had eight cookies.

| 1. Who has more cookies: Fran? **Stan?** | 2. Who has less cookies: Dan? **Nan?** |
|---|---|
| 3. Who has the most cookies: Dan? Fran? **Nan?** | 4. Who has the least cookies: Dan? Stan? **Fran?** |
| 5. Nan has how many more cookies than Fran? **8 - 2 = 6** | 6. How many cookies do Dan and Fran have together? **5 + 2 = 7** |
| 7. How many cookies do Nan and Fran have together? **8 + 2 = 10** | 8. Dan has how many more cookies than Stan? **5 - 3 = 2** |
| 9. How many cookies do Stan and Dan have together? **3 + 5 = 8** | 10. Nan has how many more cookies than Dan? **8 - 5 = 3** |

38

---

Name _____   Standard: Word Problems
Read the story then answer the questions below.

Ten squirrels were sitting in a tree. Six of the squirrels left to find some nuts. Two of the squirrels jumped to another tree. The squirrels found nine nuts. They ate the nuts for lunch.

| 1. In the beginning, how many squirrels were sitting in the tree? **10** | 2. How many squirrels first went to look for nuts? **6** |
|---|---|
| 3. How many nuts did the squirrels find? **9** | 4. How many squirrels jumped to another tree? **2** |
| 5. How many squirrels were left in the tree when six left to find nuts? **10 - 6 = 4** | 6. How many squirrels were still in the tree when two jumped to another tree? **4 - 2 = 2** |
| 7. How many squirrels left the tree all together? **6 + 2 = 6** | 8. How many more squirrels went to look for nuts than jumped to another tree? **6 - 2 = 4** |
| 9. What did the squirrels do with the nuts? **they ate the nuts** | 10. There were how many more squirrels than nuts? **10 - 9 = 1** |

39

---

Name _____   Standard: Word Problems
Read the story then answer the questions below.

Cara has a lot of books in her bedroom. She has six books about cats, three books about dogs, seven books about birds, and eleven books about cooking.

| 1. What is the name of the girl in this problem? **Cara** | 2. How many books about cats does Cara have? **6** |
|---|---|
| 3. Cara has how many books about dogs? **3** | 4. How many books does she have that are about birds? **7** |
| 5. She has how many books about cooking? **11** | 6. How many books about cats and dogs together? **6 + 3 = 9** |
| 7. How many of Cara's books are about animals? **6 + 3 + 7 = 16** | 8. Cara has how many more books about birds than dogs? **7 - 3 = 4** |
| 9. How many more books are about cooking than birds? **11 - 7 = 4** | 10. How many books does Cara have in all? **6 + 3 + 7 + 11 = 27** |
| 11. Where does Cara keep her books? **in her bedroom** | 12. Which type of book is Cara's favorite? **the story doesn't say** |

40

---

Name _____   Standard: Word Problems
Read the story then answer the questions below.

Liam went to the store for some candy. He walked three blocks west and two blocks north. He bought a lollipop for ten cents, a bag of chips for twenty-five cents, and a pack of gum for fifteen cents.

| 1. What is the name of the boy in this problem? **Liam** | 2. How many blocks did Liam walk in all? **3 + 2 = 5** |
|---|---|
| 3. How many blocks to the south did Liam walk? **none (0)** | 4. What three things did Liam buy at the store? **lollipop, chips, gum** |
| 5. How much did the bag of chips cost? **25 ¢** | 6. How much did the chips and lollipop cost together? **10 + 25 = 35 ¢** |
| 7. Liam walked how many more blocks west than north? **3 - 2 = 1** | 8. What was the name of the store? **the story doesn't say** |
| 9. How much more did the chips cost than the lollipop? **25 - 10 = 15 ¢** | 10. Which item cost Liam the least amount of money? **lollipop** |
| 11. How much money did Liam spend for these three items? **10 + 25 + 15 = 50 ¢** | 12. Which item cost Liam the most money? **chips** |

41

**ANSWER KEYS: 42, 43, 44, 45**

---

Name _____   Standard: Word Problems

Read the story then answer the questions below.

It rained four inches on Monday, two inches on Tuesday, six inches on Wednesday, and one inch on Thursday. Friday and Saturday the sun was shining and it did not rain a drop!

| 1. How much rain fell on Thursday?  **1 inch** | 2. On which day did it rain the most?  **Wednesday** |
|---|---|
| 3. How much more rain fell Wednesday than Monday?  **6 - 4 = 2 inches** | 4. How much rain fell on Monday and Friday together?  **4 + 0 = 4 inches** |
| 5. How much rain fell all together?  **4 + 2 + 6 + 1 = 13 in.** | 6. On which days didn't it rain at all?  **Friday and Saturday** |
| 7. What was the weather like on Sunday?  **the story doesn't say** | 8. How much rain fell on Monday, Tuesday and Thursday?  **4 + 2 + 1 = 7 in.** |
| 9. Which two days did it rain most?  **Monday and Tuesday** / **Wednesday and Thursday** | 10. How much rain fell on the days starting with the letter "T"?  **2 + 1 = 3 in.** |
| 11. On how many days did it rain?  **4** | 12. On which rainy day did the least amount of rain fall?  **Thursday** |

42

---

Name _____   Standard: Word Problems

Read the story then answer the questions below.

Sloan had one quarter, two dimes, one nickel, and seven pennies. He bought two pencils for sixteen cents. He paid eight more cents to get a sticker. Now Sloan wants to buy a book that costs twenty-nine cents.

| 1. How many pennies did Sloan have at first?  **7 pennies** | 2. Where did Sloan buy these items?  **the story doesn't say** |
|---|---|
| 3. How many pencils did Sloan buy?  **2 pencils** | 4. How much money did Sloan have at first?  **25 + 20 + 5 + 7 = 57 ¢** |
| 5. How much did Sloan have left after he bought the pencils?  **57 - 16 = 41 ¢** | 6. How much money did Sloan spend on pencils and stickers?  **16 + 8 = 24** |
| 7. How much does the book cost?  **29 ¢** | 8. Does Sloan have enough money left to buy the book?  **yes** |
| 9. What is the book about?  **the story doesn't say** | 10. Which is worth more?  **two dimes and seven pennies** / **one quarter** |
| 11. Which cost more, the pencils or the sticker?  **the pencils** | 12. How much did the sticker cost Sloan?  **8 ¢** |

43

---

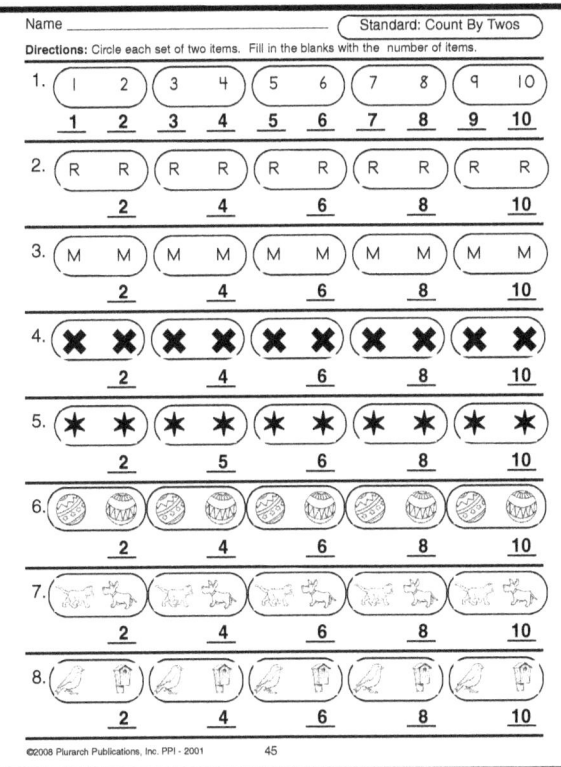

44

45

# ANSWER KEYS: 46, 47, 48, 49

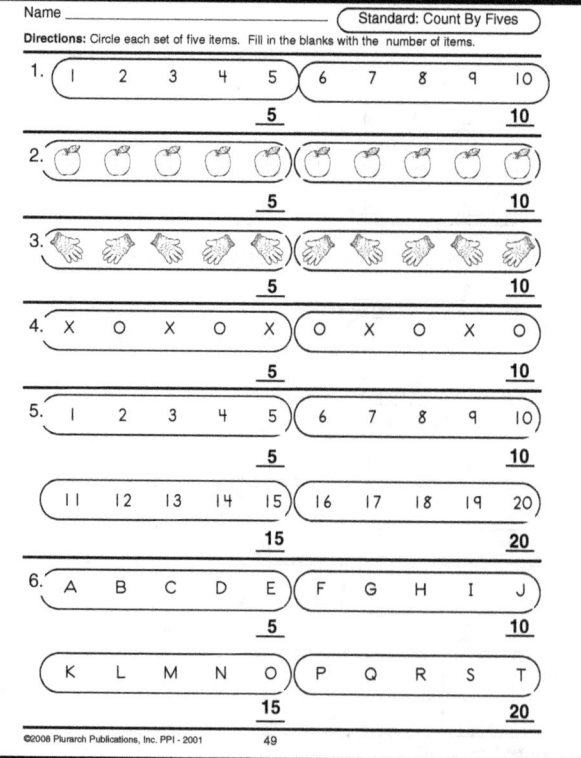

**ANSWER KEYS: 50, 51, 52, 53**

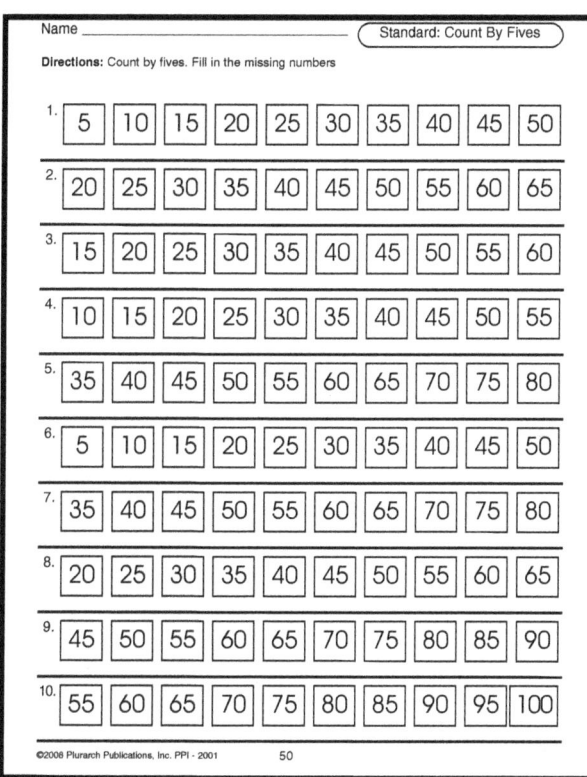

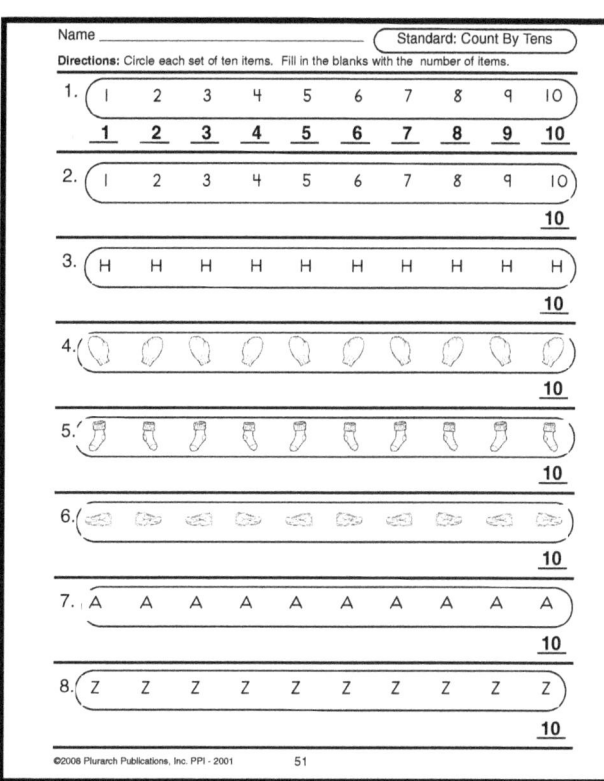

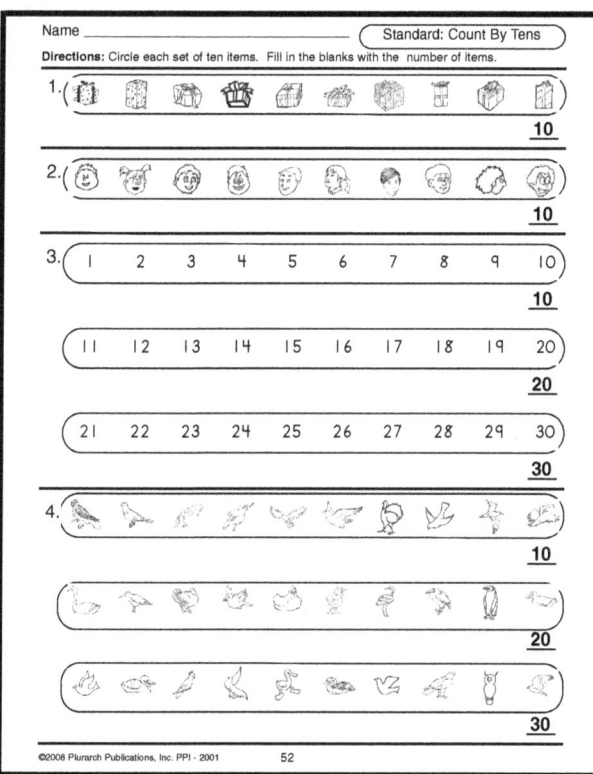

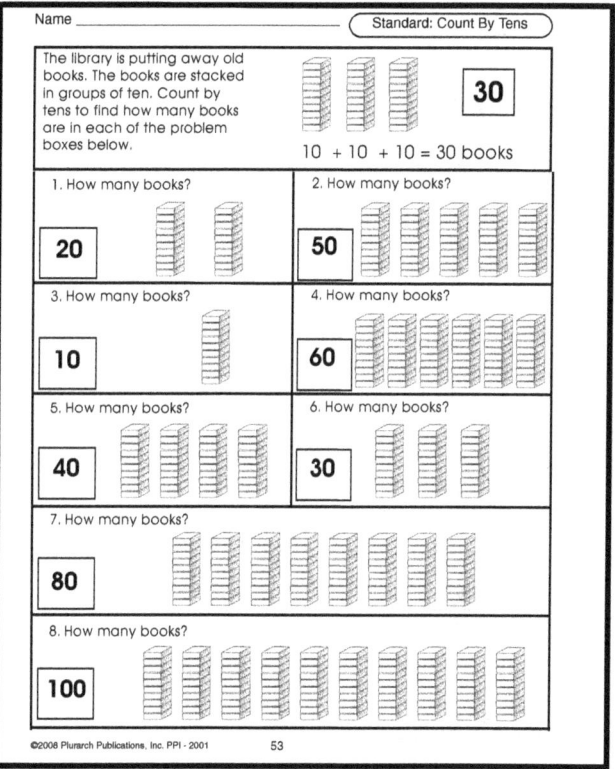

# ANSWER KEYS: 54, 55, 56, 57

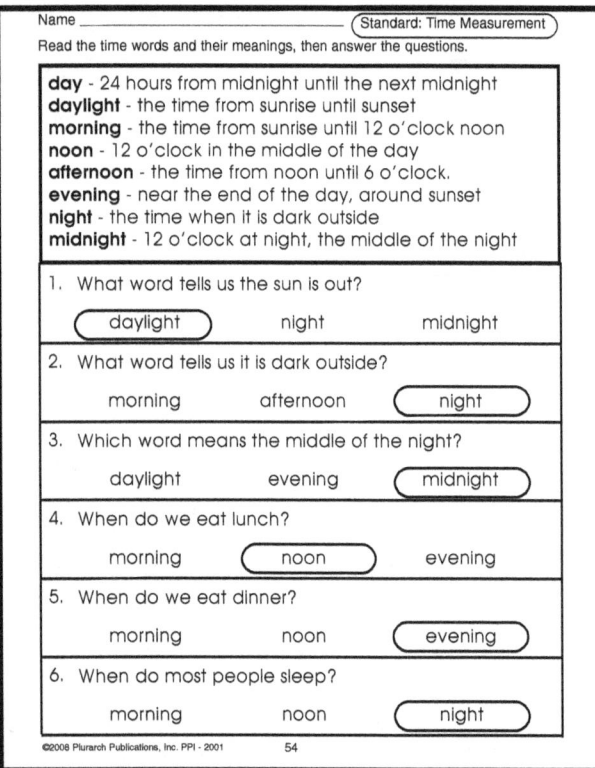

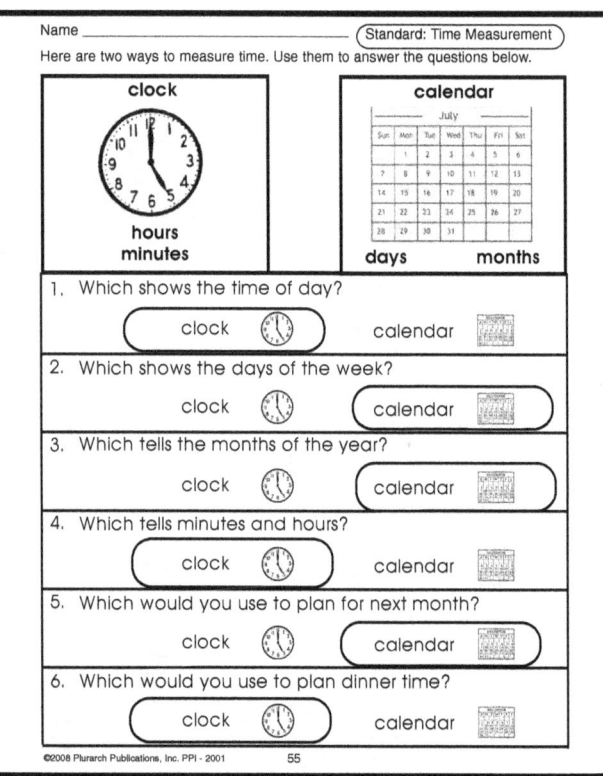

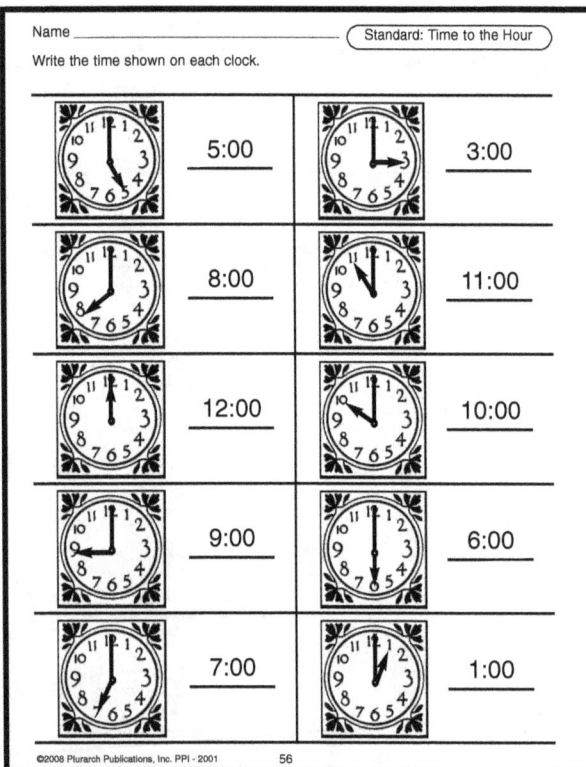

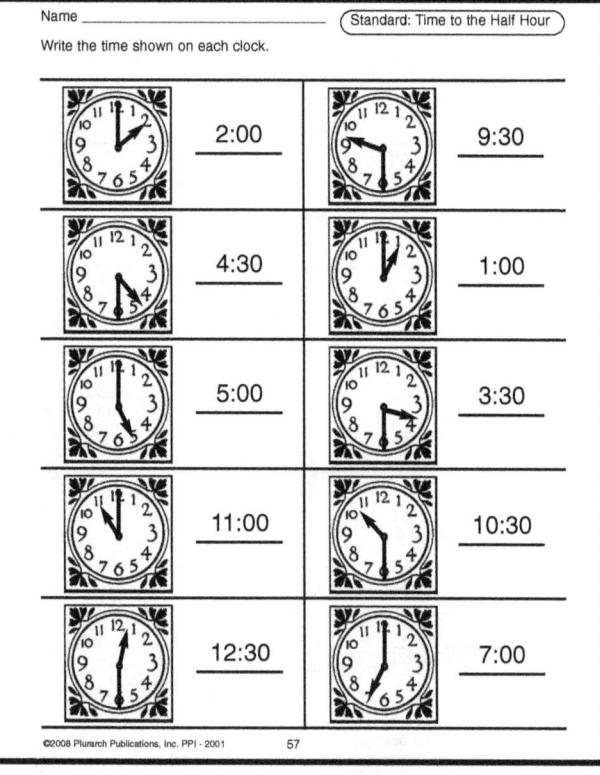

ANSWER KEYS: 58, 59, 60. 61

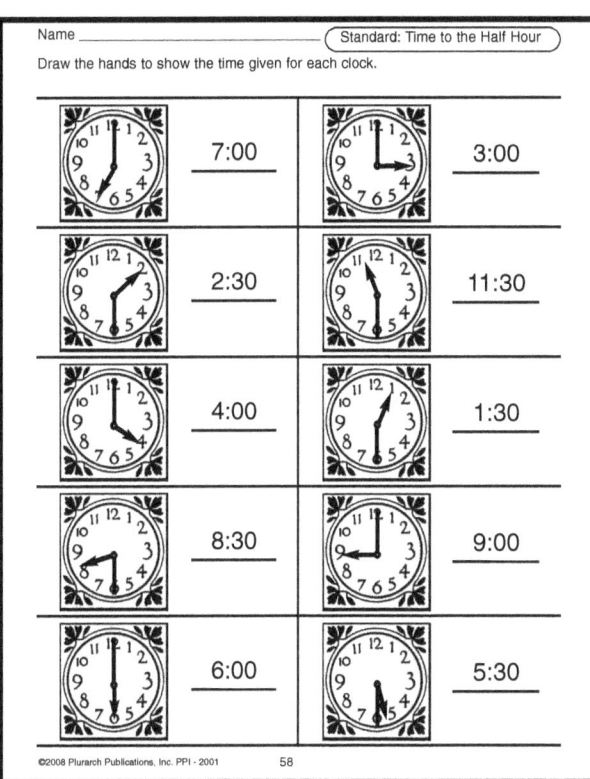

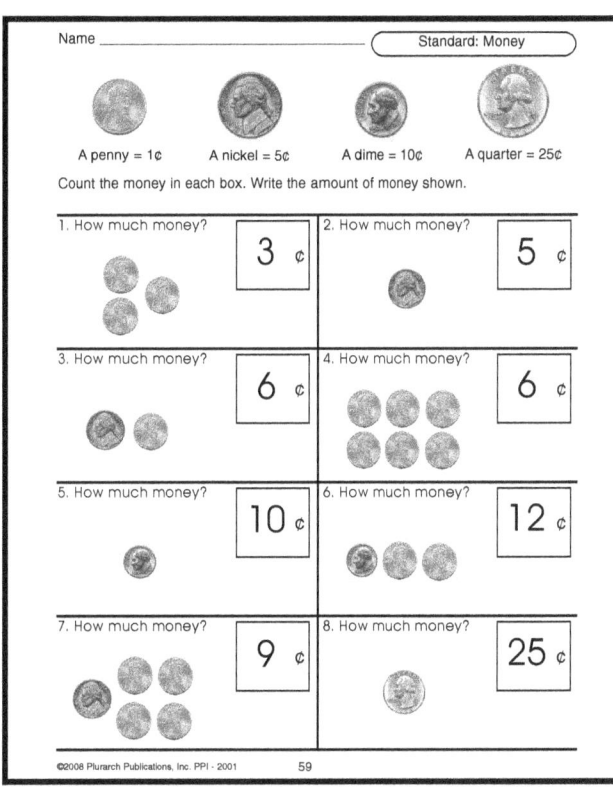

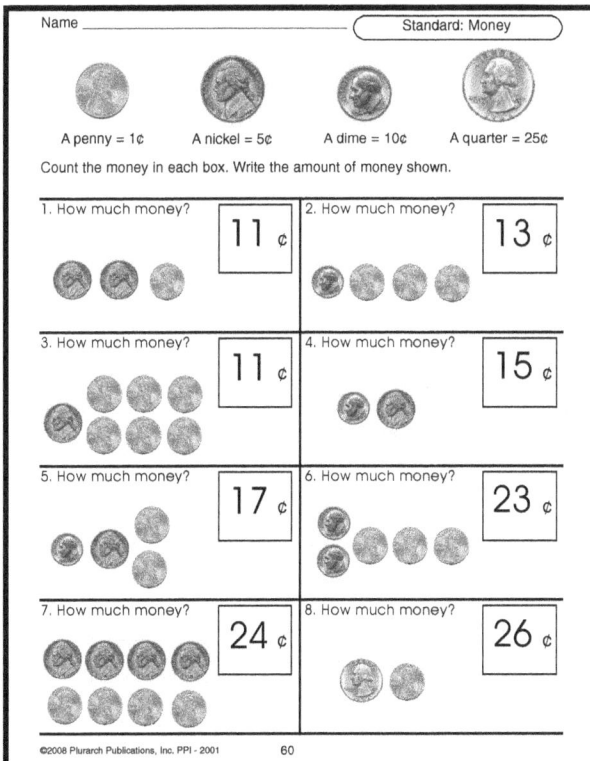

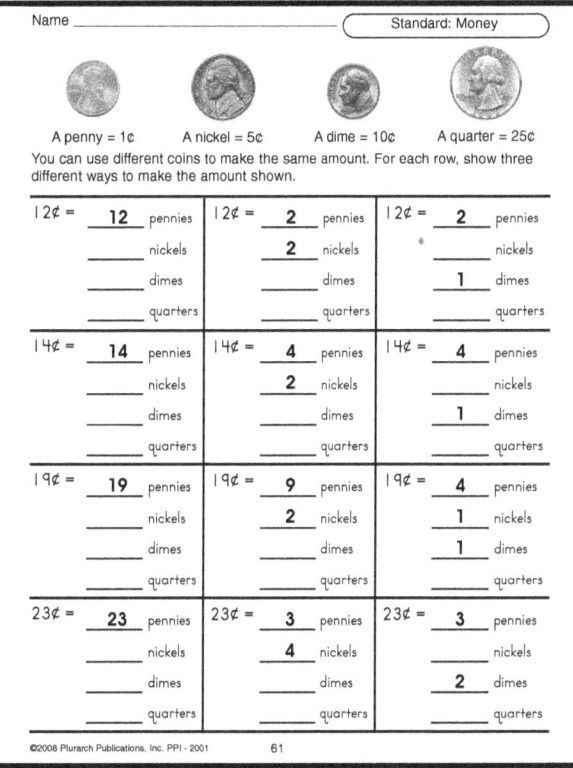

83

# ANSWER KEYS: 62, 63, 64, 65

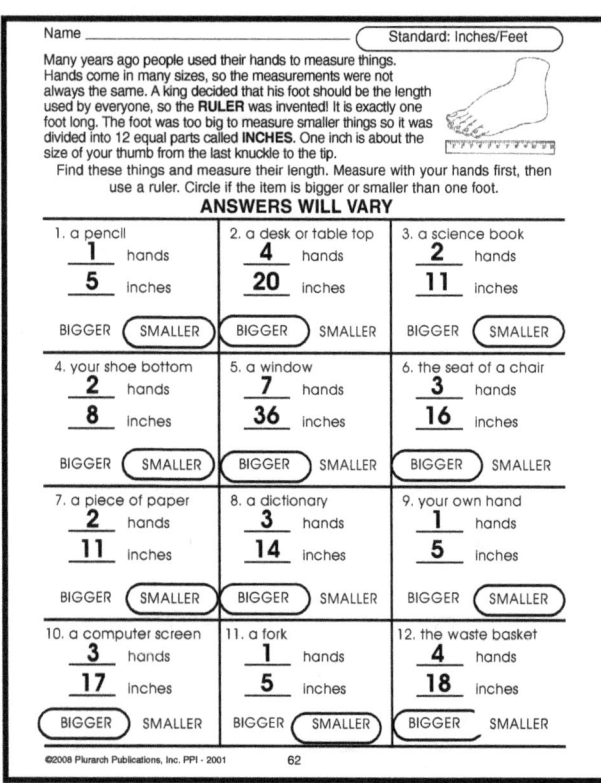

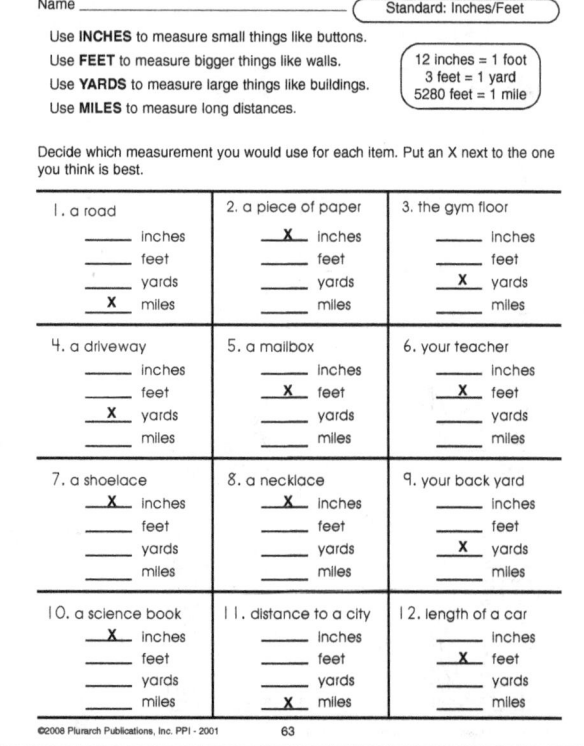

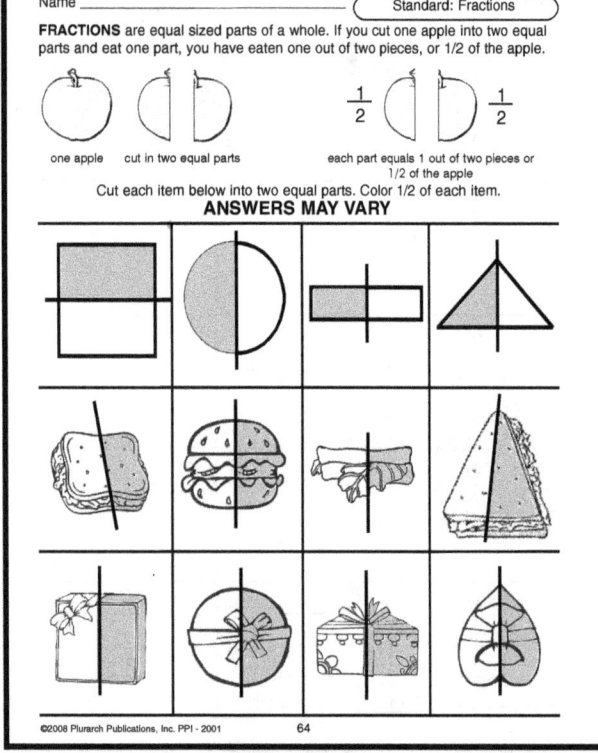

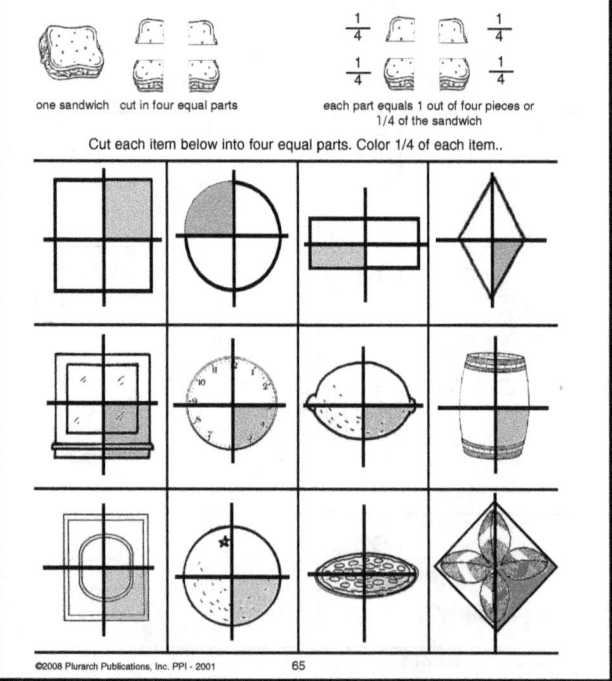

## ANSWER KEYS: 66, 67, 68, 69

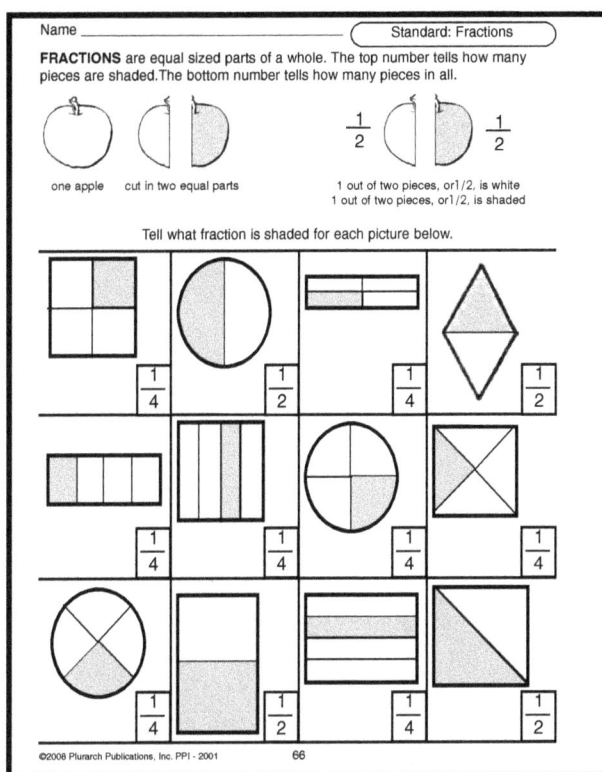

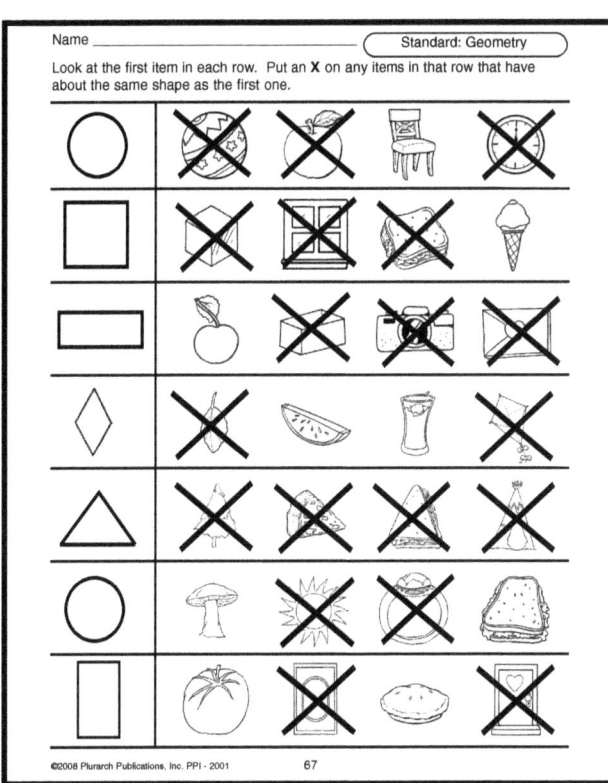

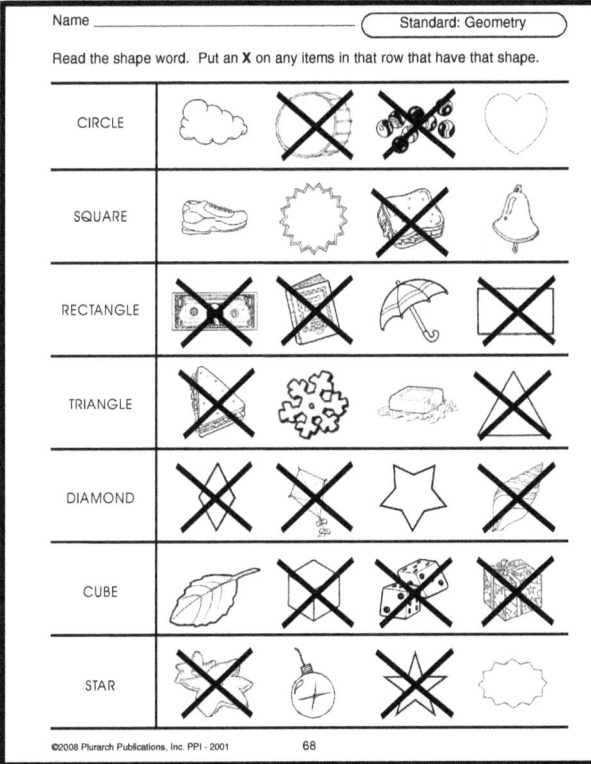

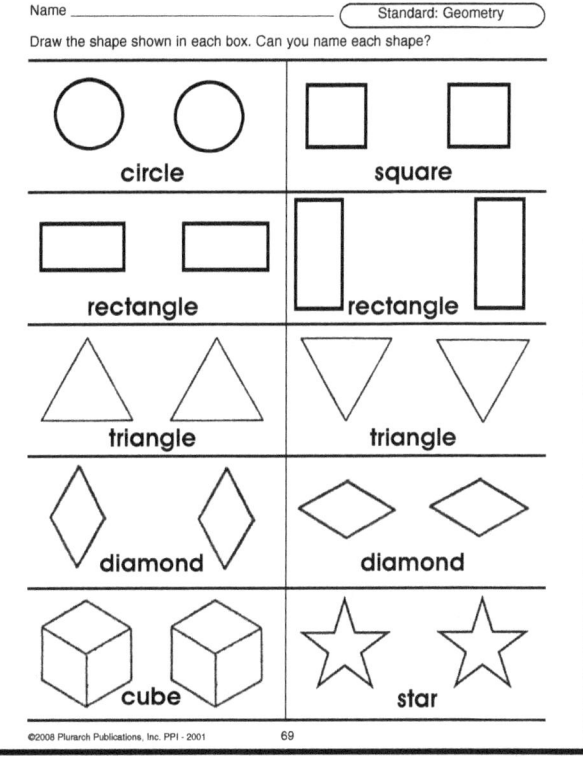

www.ingramcontent.com/pod-product-compliance
Lightning Source LLC
Chambersburg PA
CBHW081455060426
42444CB00037BA/3253